Interdisciplinary Statistics

CORRESPONDENCE ANALYSIS in PRACTICE

Second Edition

CHAPMAN & HALL/CRC
Interdisciplinary Statistics Series

Series editors: N. Keiding, B. Morgan, T. Speed, P. van der Heijden

Interdisciplinary Statistics

CORRESPONDENCE ANALYSIS in PRACTICE

Second Edition

Michael Greenacre

Universitat Pompeu Fabra

Barcelona, Spain

Chapman & Hall/CRC
Taylor & Francis Group

Boca Raton London New York

Chapman & Hall/CRC is an imprint of the
Taylor & Francis Group, an **informa** business

First edition published by Academic Press in 1993.

Chapman & Hall/CRC
Taylor & Francis Group
6000 Broken Sound Parkway NW, Suite 300
Boca Raton, FL 33487-2742

© 2007 by Taylor & Francis Group, LLC
Chapman & Hall/CRC is an imprint of Taylor & Francis Group, an Informa business

No claim to original U.S. Government works
Printed in the United States of America on acid-free paper
10 9 8 7 6 5 4 3 2 1

International Standard Book Number-10: 1-58488-616-1 (Hardcover)
International Standard Book Number-13: 978-1-58488-616-7 (Hardcover)

Visit the Taylor & Francis Web site at
http://www.taylorandfrancis.com

and the CRC Press Web site at
http://www.crcpress.com

To Françoise, Karolien and Gloudina

Contents

Preface

This book is the totally revised and extended edition of *Correspondence Analysis in Practice*, first published in 1993. In the first edition I wrote the following in the Preface, which is still relevant today:

> Correspondence analysis is a statistical technique which is useful to all students, researchers and professionals who collect categorical data, for example data collected in social surveys. The method is particularly helpful in analysing cross-tabular data in the form of numerical frequencies, and results in an elegant but simple graphical display which permits more rapid interpretation and understanding of the data. Although the theoretical origins of the technique can be traced back over 50 years, the real impetus to the modern application of correspondence analysis was given by the French linguist and data analyst Jean-Paul Benzécri and his colleagues and students, working initially at the University of Rennes in the early 1960s and subsequently at the Jussieu campus of the University of Paris. Parallel developments of correspondence analysis have taken place in the Netherlands and Japan, centred around such pioneering researchers as Jan de Leeuw and Chikio Hayashi. My own involvement with correspondence analysis commenced in 1973 when I started my doctorate in Benzécri's Laboratory of Data Analysis in Paris. The publication of my first book *Theory and Applications of Correspondence Analysis* in 1984 coincided with the beginning of a wider dissemination of correspondence analysis outside of France. At that time I expressed the hope that my book would serve as a springboard for a much wider and more routine application of correspondence analysis in the future. The subsequent evolution and growing popularity of the method could not have been more gratifying, as hundreds of researchers were introduced to the method and became familiar with its ability to communicate complex tables of numerical data to non-specialists throught the medium of graphics. Researchers with whom I have communicated come from such varying backgrounds as sociology, ecology, paleontology, archeology, geology, education, medicine, biochemistry, microbiology, linguistics, marketing research, advertising, religious studies, philosophy, art and music. ... In 1989 I was invited by Jay Magidson of Statistical Innovations Inc. to collaborate with Leo Goodman and Clifford Clogg in the presentation of a two-day short course in New York, entitled "Correspondence Analysis and Association Models: Geometric Representation and Beyond". The participants were mostly marketing professionals from major American companies. For this course I prepared a set of notes which reinforced the practical, user-oriented approach to correspondence analysis. ... The positive reaction of the audience was infectious and inspired me subsequently to present short courses on correspondence analysis in South Africa, England and Germany. It is from the notes prepared from these courses that this book has grown.

Extract from Preface of First Edition of Correspondence Analysis in Practice

In 1991 Prof. Walter Kristof of Hamburg University proposed that we organize a conference on correspondence analysis, with the assistance of Jörg Blasius of the *Zentralarchiv für Empirische Sozialforschung* (Central Archive for Empirical Social Research) at the University of Cologne. This conference was the

The Cologne and Barcelona conferences

first international one of its kind and drew a large audience to Cologne from Germany and neighbouring European countries. This initial meeting developed into a series of quadrennial conferences, repeated in 1995 and 1999 in Cologne, in 2003 at the Pompeu Fabra University in Barcelona and due to take place in 2007 at the Erasmus University in Rotterdam. The 1991 conference led to the publication of the book *Correspondence Analysis in the Social Sciences*, while the 1995 conference gave birth to another book *Visualization of Categorical Data*, both of which received excellent reviews. For the 1999 conference on *Large Scale Data Analysis* participants had to present analyses of data from the multinational International Social Survey Programme (ISSP). This interdisciplinary meeting included presentations not only on the latest methodological developments in survey data analysis, but also topics as diverse as religion, the environment and social inequality. In 2003 we returned to the original theme for the Barcelona conference, which was baptized with the Catalan girl's name CARME (Correspondence Analysis and Related MEthods); hence the formation of the CARME network (www.carme-n.org). This led to Jörg Blasius and myself editing a third book, *Multiple Correspondence Analysis and Related Methods*, which was published by Chapman & Hall in June 2006. As with the two previous volumes, our idea was to produce a multi-authored book, inviting experts in the field to contribute, with our task being to write the introductory and linking material, unifying the notation and compiling a common reference list and index. These books mark, in some sense, the pace of development of the subject, at least in the social sciences, and are highly recommended to anyone interested in deepening their knowledge on correspondence analysis and methods related to it.

New material in Second Edition I have been very gratified to rewrite *Correspondence Analysis in Practice*, having accumulated considerably more experience in social and environmental research in the 13 years since the publication of the first edition. I have given more short courses across the world, including Spain, Italy, Belgium, Brazil, Canada and Australia, and also courses on multivariate analysis for environmental biologists in Norway, Iceland and Spain, with correspondence analysis as one of the major topics. The experience of giving these courses has provided new applications and new ideas for this new edition, so that I have been able to bring the material right up to date. Apart from completely revising the original chapters and changing several examples, five new chapters have been added, on "Transition and Regression Relationships" (showing how the results and the data are linked by linear functions), "Stacked Tables" (showing how several cross-tables can be analysed jointly), "Subset Correspondence Analysis" (a simple but highly effective variation of the correspondence analysis algorithm to analyse selected parts of a data set), "Analysis of Square Tables" (showing how square tables — e.g., social mobility tables, brand-switching matrices — can be decomposed into parts that can be visualized separately using correspondence analysis), and "Canonical Correspondence Analysis" (showing how to take into account relationships to external predictor variables, an extension of correspondence analysis that is very popular in ecology). All in all, I can say that this second edition contains almost all my practical knowledge of the subject, after 33 years working in this area.

At a conference I attended in the 1980s, I was given this lapel button with its nicely ambiguous maxim, which could well be the motto of correspondence analysts all over the world: *Comparison of first and second editions*

To illustrate the more obvious meaning of this motto, and to give a simple example of correspondence analysis, I counted how many tables and figures there were in each of the chapters of the first edition and in their counterpart chapters in the second edition. For a valid comparison between the two editions I expressed these relative to the total number of pages in each chapter and then performed a variation of the technique called "subset correspondence analysis" (described in Chapter 21), using the standard biplot presentation (described in Chapter 13), leading to the data map shown on the following page (you will understand these ideas after you read the book, but for the moment think of this example as a type of scatterplot). The two vectors *Figures* and *Tables* point to the right, so chapters to the right have higher than average percentage of figures and tables. For example, Chapters 9 and 11 in the second edition have the highest percentage of figures and tables, respectively. In fact, about three quarters of the second edition's chapters represented in the map are to the right of the centre of the map (the centre represents the average), with only one quarter of the first edition's chapters to the right, demonstrating the substantial increase in figures and tables in the second edition. In fact, the information content and applications component of the book have increased substantially compared to the first edition, which had a main body (excluding appendices) of 177 pages — in the second edition this has increased to 200 pages, an increase of 13%, while there has been an increase of 25% in chapters and topics, 18% in data sets, 44% in figures and 63% in tables.

This book has some innovations that deserve mentioning. Like the first edition, it is intended to be didactic, but this edition is even more so. I wanted each chapter to represent a fixed amount to read or teach, and there was no better way to do that than to limit the length of each chapter — each chapter is exactly eight pages in length. This was one of the most interesting aspects of *Format of second edition*

Exhibit 0.1:
Subset correspondence analysis map of percentages of figures and tables in 20 comparable chapters from first and second editions, presented as a standard biplot. Only chapters of the present second edition (denoted by B) that have a counterpart in the first edition (denoted by A) are analysed. The chapter numbers correspond to the present edition, giving the same numbers to the corresponding chapters in the first edition.

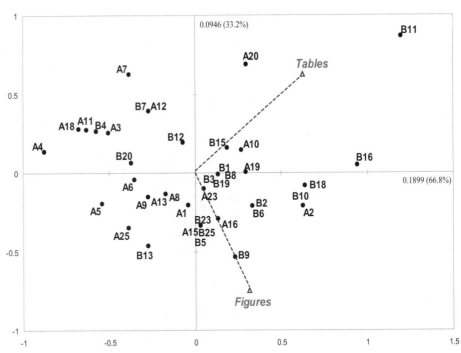

the project and evolved naturally as I was writing it. One of my colleagues remarked that it was like writing 14-line sonnets with strict rules for rhyming, which was certainly true in my case: the format definitely contributed to the creative process. Another innovation was the extensive use of marginal notes. Every paragraph of the book has a heading which is placed in the margin. Then to summarize each chapter, these paragraph headings are all gathered into a "Contents" section on the first page of every chapter. All figure and table captions are also placed in the margins, so that the captions are often much longer and more informative than conventional captions that tend to be one-liners. Finally, each chapter ends with a summary in the form of a list.

Theoretical appendix As in the first edition, the book's main thrust is still toward the practice of correspondence analysis, so most technical issues and mathematical aspects are gathered in an Appendix at the end of the book. This theoretical appendix is more extensive than that of the first edition, including additional theory on the new topics in the book such as canonical correspondence analysis.

Computational One of the main features of this edition which distinguishes it from the original
appendix one, and which clearly marks the present digital era, is the lengthy computational appendix, using the freely available R program, which has become the *de facto* standard in statistical computing. Almost all the analyses in the

book are referred to in this appendix and the R commands are given to obtain the corresponding results. In addition, I describe three different technologies that I used to create the graphical displays in the book. As simple as the data maps might seem, constructing them is by no means a trivial exercise.

No references at all are given in the 25 chapters, and a relatively brief bibliographical appendix is given to point readers toward further readings that include much more complete literature guides. A glossary of the most important terms and an epilogue with some final thoughts conclude the book. *Bibliographical appendix, glossary and epilogue*

Acknowledgements

The first edition of this book was written in South Africa, and the present edition in Catalonia, Spain. Many people and institutions have contributed in one way or another to this project. First and foremost, I would like to thank Rafael Pardo, director of the *Fundación BBVA* in Madrid — without the support of this foundation and its director, both in encouragement and in financial support to enable me to take six months' leave from my teaching, I can honestly say that this book would not be in your hands now. In addition, the BBVA Foundation plans to publish the Spanish translation of this book. Then there is the *Universitat Pompeu Fabra* in Barcelona, where I have been working since 1994 — I express my thanks to the director Xavier Calsamiglia and the whole department for giving me the liberty to devote so much of my time to this project.

I would like to thank all my friends and colleagues in many countries for moral and intellectual support, especially Zerrin Aşan, Jörg Blasius, John Gower, Carles Cuadras, Trevor Hastie, Michael Browne, Victor Thiessen, Karl Jöreskog, Lesley Andres, John Aitchison, Paul Lewi, Patrick Groenen, Pieter Kroonenberg, Ludovic Lebart, Michael Friendly, Antoine de Falguerolles, Salve Dahle, Stig Falk-Petersen, Raul Primicerio, Johs Hjellbrekke, Tom Backer Johnsen, Tor Korneliussen, Ümit Senesen, Brian Monteith, Ken Reed, Gillian Heller, Antonella Curci, Gianna Mastrorilli, Paola Bordandini, Walter Zucchini, Oleg Nenadić, Thierry Fahmy, Tamara Djermanovic, Volker Hooyberg, Gurdeep Stephens, Rita Lugli, Danilo Guaitoli and the whole community of Gréixer — you have all played a part in this story! Particular thanks go to Jörg Blasius for a thorough proofreading of the manuscript, and to Oleg Nenadić for his collaboration in preparing the **ca** package in R. Like the first edition, I have dedicated this book to my three daughters, who never cease to amaze me by their joy, sense of humour and diversity. Finally, I thank Waseem Andrabi and Shashi Kumar of International Typesetting and Composition in India, who helped in the design of the LaTeX style file that I used to typeset the book, and commissioning editor Rob Calver, production coordinator Marsha Pronin and project editor Mimi Williams of Chapman & Hall/CRC Press for placing their trust in me and for their constant cooperation in making this second edition become a reality.

Michael Greenacre
Barcelona

Scatterplots and Maps

Correspondence analysis is a method of data analysis for representing tabular data graphically. Correspondence analysis is a generalization of a simple graphical concept with which we are all familiar, namely the *scatterplot*. The scatterplot is the representation of data as a set of points with respect to two perpendicular coordinate axes: the horizontal axis often referred to as the x-axis and the vertical one as the y-axis. As a gentle introduction to the subject of correspondence analysis, it is convenient to reflect for a short time on our perception of scatterplots and how we interpret them in relation to the data they represent graphically. Particular emphasis will be placed on how we interpret distances between points in a scatterplot and when scatterplots can be seen as a spatial map of the data.

Contents

Data set 1: My travels in 2005

As I started writing this book at the end of 2005, I was thinking about the journeys I had made during the year to three of my favourite countries: Norway, Canada and Greece. According to my diary I spent 18 days in Norway, 15 days in Canada and 29 days in Greece. Apart from these trips I also made several short trips to France and Germany, totalling 24 days. This numerical description of my time spent in foreign countries can be represented graphically as in the graphs given in Exhibit 1.1. This seemingly trivial example conceals several issues which are relevant to our perception of graphs of this type that represent data with respect to two coordinate axes, and which will

Exhibit 1.1:
*Graphs of number of
days spent in foreign
countries in 2005, in
scatterplot and
bar-chart formats
respectively. A
percentage scale,
expressing days
relative to the total
of 86, is given on the
right-hand side of
each graph.*

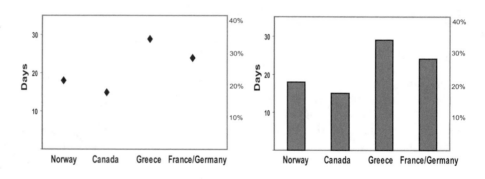

eventually help us to understand correspondence analysis. Let me highlight these issues one at a time.

*Continuous
variables*
The left-hand vertical axis labelled Days represents the scale of a numeric piece of information often referred to as a *continuous* variable. The scale on this axis is the number of days spent in some foreign country, and the ordering from zero days at the bottom end of the scale to 30 days at the top end is clearly defined. In the more common bar-chart form of this display, given in the right-hand graph of Exhibit 1.1, bars are drawn with lengths proportional to the values of the variable. Of course, the number of days is a rounded approximation of the time actually spent in each country, but we call this variable continuous because the underlying time variable is indeed truly continuous.

*Expressing
data in relative
amounts*
The right-hand vertical axis of each plot in Exhibit 1.1 can be used to read the corresponding percentage of days relative to the total of 86 days. For example, the 18 days in Norway account for 21% of the total time. The total of 86 is often called the *base* relative to which the data are expressed. In this case there is only one set of data and therefore just one base, so in these plots the original absolute scale on the left and the relative scale on the right can be depicted on the same graph.

*Categorical
variables*
In contrast to the vertical y-axis, the horizontal x-axis is clearly not a numeric variable. The four points along this axis are just positions where we have placed labels denoting the countries visited. The horizontal scale represents what is called a *categorical* variable. There are two features of this horizontal axis that have no substantive meaning in the graph: the ordering of the categories and the distances between them.

Firstly, there is no strong reason why Norway has been placed first, Canada second and Greece third, except perhaps that I visited these countries in that order. Because the France/Germany label refers to a collection of shorter trips scattered throughout the year, it was placed after the others. By the way, in this type of representation where order is essentially irrelevant, it is usually a good idea to re-order the categories in a way that has some substantive meaning, for example in terms of the values of the variable. In this example we could order the countries in descending order of days, in which case we would position the countries in the order Greece, France/Germany, Norway and Canada, from most visited to least. This simple re-arrangement assists in the interpretation of data, especially if the data set is much larger: for example, if I had visited 20 different countries, then the order would contain relevant information that is not quickly deduced from the data in their original ordering.

Secondly, there is no reason why the four points are at equal intervals apart on the axis. There is also no reason to put them at different intervals apart, so it is purely for convenience and aesthetics that they have been equally spaced. Using correspondence analysis we will show that there are substantively interesting ways to define intervals between the categories of a variable such as this one. In fact, correspondence analysis will be shown to yield quantified values for the categories where both the distances between the categories and their ordering have substantive meaning.

Since the ordering of the countries is arbitrary on the horizontal axis of Exhibit 1.1, as well as the distances between them, there would be no sense in measuring and interpreting distances between the displayed points in the left-hand graph. The only distance measurement that has meaning is in the strictly vertical direction, because of the numerical nature of the vertical axis that indicates frequency (or relative frequency).

In some special cases, the two variables that define the axes of the scatterplot are of the same numeric nature and have comparable scales. For example, suppose that 20 students have written a mathematics examination consisting of two parts, algebra and geometry, each part counting 50% towards the final grade. The 20 students can be plotted according to their pair of grades, shown in Exhibit 1.2. It is important that the two axes representing the respective grades have scales with unit intervals of identical lengths. Because of the similar nature of the two variables and their scales, it is possible to judge distances in any direction of the display, not only horizontally or vertically. Two points that are close to each other will have similar results in the examination, just like two neighbouring towns having a small distance between them. Thus, one can comment here on the shape of the scatter of points and the fact that there is a small cluster of four students with high grades and a single student with

Exhibit 1.2:
Scatterplot of grades of 20 students in two sections (algebra and geometry) of a mathematics examination. The points have spatial properties: for example, the total grade is obtained by projecting each point perpendicularly onto the 45° line, which is calibrated from 0 (bottom left corner) to 100 (top right corner).

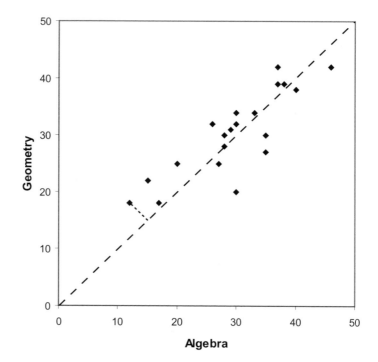

very high grades. Exhibit 1.2 can be regarded as a *map*, because the position of each student can be regarded as a two-dimensional position, almost like a geographical location in a region defined by latitude and longitude.

Calibration of a direction in the map

Maps have interesting geometric properties. For example, in Exhibit 1.2 the 45° dashed line actually defines an axis for the final grades of the students, combining the algebra and geometry grades. If this line is calibrated from 0 (bottom left) to 100 (top right), then each student's final grade can be read from the map by projecting each point perpendicularly onto this line. An example is shown of a student who received 12 out of 50 and 18 out of 50 for the two sections, respectively, and whose position projects onto the line at coordinates 15 and 15, corresponding to a total grade of 30.

Information-transforming nature of the display

The scatterplots in Exhibit 1.1 and Exhibit 1.2 are different ways of expressing in graphical form the numerical information in the two tables of travel and examination data respectively. In each case there is no loss of information between the data and the graph. Given the graph it is easy to recover the data exactly. We say that the scatterplot or map is an "information-transforming instrument" — it does not process the data at all; it simply expresses the data in a visual format that communicates the same information in an alternative way.

In my travel example, the categorical variable "country" has four categories, and, since there is no inherent ordering of the categories, we refer to this variable more specifically as a *nominal* variable. If the categories are ordered, the categorical variable is called an *ordinal* variable. For example, a day could be classified into three categories according to how much time is spent working: (i) less than one hour (which I would call a "holiday"), (ii) more than one but less than six hours (a "half day", say) and (iii) more than six hours (a "full day"). These categories, which are based on a continuous variable "time spent daily working" divided up into intervals, are ordered and this ordering is usually taken into account in any graphical display of the categories. In many social surveys, questions are answered on an ordinal scale of response, for example, an ordinal scale of importance: "not important"/"somewhat important"/"very important". Another typical example is a scale of agreement/disagreement: "strongly agree"/"somewhat agree"/"neither agree nor disagree"/"somewhat disagree"/"strongly disagree". Here the ordinal position of the category "neither agree nor disagree" might not lie between "somewhat agree" and "somewhat disagree"; for example, it might be a category used by some respondents instead of a "don't know" response when they do not understand the question or when they are confused by it. We shall treat this topic later in this book (Chapter 21) once we have developed the tools that allow us to study patterns of responses in multivariate questionnaire data.

Nominal and ordinal variables

COUNTRY	Holidays	Half Days	Full Days	TOTAL
Norway	6	1	11	18
Canada	1	3	11	15
Greece	4	25	0	29
France/Germany	2	2	20	24
TOTAL	13	31	42	86

Exhibit 1.3: *Frequencies of different types of day in four sets of trips*

Let us suppose now that the 86 days of my foreign trips were classified into one of the three categories *holidays*, *half days* and *full day*. The *cross-tabulation* of country by type of day is given in Exhibit 1.3. This table can be considered in two different ways: as a set of rows or a set of columns. For example, each column is a set of frequencies characterizing the respective type of day, while each row characterizes the respective country. Exhibit 1.4(a) shows the latter way, namely a plot of the frequencies for each country (row), where the horizontal axis now represents the type of day (column). Notice that, because the categories of the variable "type of day" are ordered, it makes sense to connect the categories by lines. Clearly, if we want to make a substantive comparison between the countries, then we should take into account the fact that differ-

Plotting more than one set of data

ent numbers of days in total were spent in each country. Each country total forms a base for the re-expression of the corresponding row in Exhibit 1.3 as a set of percentages (Exhibit 1.5). These percentages are visualized in Exhibit 1.4(b) in a plot which expresses better the different compositions of days in the respective trips.

Exhibit 1.4:
Plots of (a) frequencies in Exhibit 1.3 and (b) relative frequencies in each row expressed as percentages.

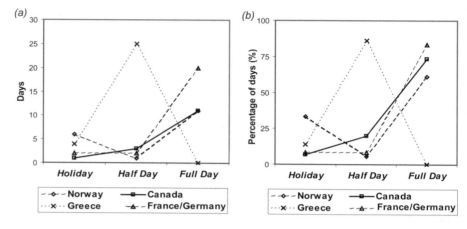

Exhibit 1.5:
Percentages of types of day in each country, as well as the percentages overall for all countries combined; rows add up to 100%.

COUNTRY	Holidays	Half Days	Full Days
Norway	33%	6%	61%
Canada	7%	20%	73%
Greece	14%	86%	0%
France/Germany	8%	8%	83%
Overall	*15%*	*36%*	*49%*

Interpreting absolute or relative frequencies

There is a lesson to be learnt from these displays that is fundamental to the analysis of frequency data. Each trip has involved a different number of days and so corresponds to a different base as far as the frequencies of the types of days is concerned. The 6 *holidays* in Norway, compared to the 4 in Greece, can be judged only in relation to the total number of days spent in these respective countries. As percentages they turn out to be quite different: 6 out of 18 is 33%, while 4 out of 29 is 14%. It is the visualization of the relative frequencies in Exhibit 1.4(b) that gives a more accurate comparison of how I spent my time in the different countries. The "marginal" frequencies (18, 15, 29, 24 for the countries, and 13, 31, 42 for the day types) are also interpreted relative to their respective totals — for example, the last row of Exhibit 1.5 shows the percentages of day types for all countries combined, and could also be plotted in Exhibit 1.4(b).

Any conclusion drawn from the points positions in Exhibit 1.4(b) is purely an interpretation of the data and not a statement of the statistical significance of the observed feature. In this book we shall address the statistical aspects of graphical displays only at the end of the book (Chapter 25); for the most part we shall be concerned only with the question of data description and interpretation. The deduction that I had proportionally more holidays in Norway than in the other countries is certainly true in the data and can be seen strikingly in Exhibit 1.4(b). It is an entirely different question whether this phenomenon is statistically compatible with a model or hypothesis of my behaviour that postulates that the proportion of holidays was the same across all trips. Most of statistical methodology concentrates on problems where data are fitted and compared to a theoretical model or preconceived hypothesis, with little attention being paid to enlightening ways for describing data, interpreting data and generating hypotheses. A typical example in the social sciences is the use of the ubiquitous chi-square statistic to test for association in a cross-tabulation. Often statistically significant association is found but there are no simple tools for detecting which parts of the table are responsible for this association. Correspondence analysis is one tool that can fill this gap, allowing the data analyst to see the pattern of association in the data and to generate hypotheses that can be tested in a subsequent stage of research. In most situations data description, interpretation and modelling can work hand-in-hand with each other. But there are situations where data description and interpretation assume supreme importance, for example when the data represent the whole population of interest.

Describing and interpreting data, vs. modelling and statistical inference

As data tables increase in size, it becomes more difficult to make simple graphical displays such as Exhibit 1.4, owing to the overabundance of points. For example, suppose I had visited 20 countries during the year and had a breakdown of time spent in each one of them, leading to a table with many more rows. I could also have recorded other data about each day in order to study possible relationships with the type of day I had; for example, the weather on each day — "fair weather", "partly cloudy", or "rainy". So the table of data might have many more columns as well as rows. In this case, to draw graphs such as Exhibit 1.4, involving many more categories and with 20 sets of points traversing the plot, would result in such a confusion of points and symbols that it would be difficult to see any patterns at all. It would then become clear that the descriptive instrument being used, the scatterplot, is inadequate in bringing out the essential features of the data. This is a convenient point to introduce the basic concepts of correspondence analysis, which is also a method for graphically describing tabular data graphically, but which can easily accommodate larger data sets.

Large data sets

SUMMARY:
Scatterplots and
Maps

1. Scatterplots involve plotting two variables, with respect to a horizontal axis and a vertical axis, often called the "x-axis" and "y-axis" respectively.

2. Usually the x variable is a completely different entity to the y variable. We can often interpret distances along at least one of the axes in the specific sense of measuring the distance according to the scale that is calibrated on the axis. It is usually meaningless to measure or interpret oblique distances in the plot.

3. In some cases the x and y variables are similar entities with comparable scales, in which case interpoint distances can be interpreted as a measure of difference, or dissimilarity, between the plotted points. In this special case we call the scatterplot a *map*.

4. When plotting positive quantities (usually frequencies in our context), both the absolute and relative values of these quantities are of interest.

5. The more complex the data are, the less convenient it is to represent these data in a scatterplot.

6. This book is concerned with describing and interpreting complex information, rather than modelling it.

Profiles and the Profile Space

The concept of a set of relative frequencies, or a *profile*, is fundamental to correspondence analysis (referred to from now on by its abbreviation CA). Such sets, or *vectors*, of relative frequencies have special geometric features because the elements of each set add up to 1 (or 100%). In analysing a frequency table we can look at the relative frequencies for rows or for columns, called row or column profiles respectively. In this chapter we shall show how profiles can be depicted as points in a profile space, illustrating the concept in the special case when the profile consists of only three elements.

Contents

Let us look again at the data in Exhibit 1.3, a table of frequencies with four rows (the countries) and three columns (the type of day). The first and most basic concept in CA is that of a *profile*, which is a set of frequencies divided by their total. Exhibit 2.1 shows the row profiles for these data: for example, the profile of Norway is [0.33 0.06 0.61], where $0.33 = 6/18, 0.06 = 1/18, 0.61 = 11/18$. We say that this is the "profile of Norway across the types of day". The profile may also be expressed in percentage form, i.e. [33% 6% 61%] in this case, as in Exhibit 1.5. In a similar fashion the profile of Canada across the day types is [0.07 0.20 0.73], concentrated in the *full day* category, as is Norway. In contrast, Greece has a profile of [0.14 0.86 0.00], concentrated in the *half day* category, and so on. These are the values plotted in Exhibit 1.4(b) on page 6.

Profiles

COUNTRY	Holidays	Half Days	Full Days
Norway	0.33	0.06	0.61
Canada	0.07	0.20	0.73
Greece	0.14	0.86	0.00
France/Germany	0.08	0.08	0.83
Average	*0.15*	*0.36*	*0.49*

Average profile   In addition to the four country profiles, there is an additional row in Exhibit 2.1 labelled *Average*. This is the profile of the final row [13 31 42] of Exhibit 1.3, which contains the column sums of the table; in other words this is the profile of all the trips put together. In Chapter 3 we shall explain more specifically why this is called the *average profile*. For the moment, it is only necessary to realize that, out of the total of 86 days travelled, irrespective of country visited, 15% were *holidays*, 36% were *half days* and 49% were *full days* of work. When comparing profiles we can compare one country's profile with another, or we can compare a country's profile with the average profile. For example, eyeballing the figures in Exhibit 2.1, we can see that of all the countries, the profiles of Canada and France/Germany are the most similar. Compared to the average profile, these two profiles have a higher percentage of *full days* and are below average on *holidays* and *half days*.

*Row profiles
and column
profiles*   In the above we looked at the row profiles in order to compare the different countries. We could also consider Exhibit 1.3 as a set of columns and compare how the different types of days are distributed across the countries. Exhibit 2.2 shows the column profiles as well as the average column profile. For example, of the 13 *holidays* 46% were in Norway, 8% in Canada, 31% in Greece and 15% in France/Germany, and so on for the other columns. Since I spent different numbers of days in each country, these figures should be checked against those of the average column profile to see whether they are lower or higher than the average pattern. For example, 46% of all *holidays* were spent in Norway, whereas the number of days spent in Norway was just 21% of the total of 86 — in this sense there is a high number of *holidays* there compared to the average.

COUNTRY	Holiday	Half Day	Full Day	Average
Norway	0.46	0.03	0.26	*0.21*
Canada	0.08	0.10	0.26	*0.17*
Greece	0.31	0.81	0.00	*0.34*
France/Germany	0.15	0.07	0.48	*0.28*

Looking again at the proportion 0.46 (= 6/13) of *holidays* spent in Norway (Exhibit 2.2) and comparing it to the proportion 0.21 (= 18/86) of all days spent in that country, we could calculate the ratio 0.46/0.21 = 2.2, and conclude that *holidays* in Norway were just over twice the average. We come to exactly the same conclusion if we make a similar calculation on the row profiles. In Exhibit 2.1 the proportion of *holidays* in Norway was 0.33 (= 6/18) whereas for all countries the proportion was 0.15 (= 13/86). Thus, there are 0.33/0.15 = 2.2 times as many compared to the average, the same ratio as we obtained when arguing from the point of view of the column profiles (this ratio is called the *contingency ratio* and will re-appear in future chapters). Whether we argue via the row profiles or column profiles we arrive at the same conclusion. In Chapter 8 it will be shown that CA treats the rows and columns of a table in an equivalent fashion or, as we say, in a *symmetric* way.

Symmetric treatment of rows and columns

Nevertheless, it is true in practice that a table of data is often thought of and interpreted in a non-symmetric, or *asymmetric*, fashion, either as a set of rows or as a set of columns. For example, since each row of Exhibit 1.3 constitutes a different journey, it might be more natural to think of the table row-wise, as in Exhibit 2.1. Deciding which way is more appropriate depends on the nature of the data and the researcher's objective, and the decision is often not a conscious one. One concrete manifestation of the actual choice is whether the researcher refers to row or column percentages when interpreting the data. Whatever the decision, the results of CA will be invariant to this choice.

Asymmetric consideration of the data table

Let us consider the four row profiles and average profile in Exhibit 2.1 and a completely different way to plot them. Rather than the display of Exhibit 1.4(b), where the horizontal axis serves only as labels for the type of day and the vertical axis represents the percentages, we now propose using three axes corresponding to the three types of day, like a scatterplot in three dimensions. To imagine three perpendicular axes is not difficult: merely look down into an empty corner of the room you are sitting in and you will see three axes as shown in Exhibit 2.3. Each of the three edges of the room serves as an axis for plotting the three elements of the profile. These three values are now considered to be coordinates of a single point that represents the whole profile — this is quite different from the graph in Exhibit 1.4(b) where there is a point for each of the three profile elements. The three axes are labelled *holidays*, *half days* and *full days*, and are calibrated in fractional profile units from 0 to 1. To plot the four profiles is now a simple exercise. Norway's profile of [0.33 0.06 0.61] (see Exhibit 2.1) is 0.33 of a unit along axis *holidays*, 0.06 along axis *half days* and 0.61 along *full days*. To take another example, Greece's profile of [0.14 0.86 0.00] has a zero coordinate in the *full days* direction, so its position is on the "wall", as it were, on the left-hand side of the display, with coordinates 0.14 and 0.86 on the two axes *holidays* and *half days* defining

Plotting the profiles in the profile space

Exhibit 2.3:
*Positions of the four
row profiles (•) of
Exhibit 2.1 as well
as their average
profile (∗)in
three-dimensional
space, depicted as
the corner of a room
with "floor tiles".
For example,
Norway is 0.06 along
the half days axis,
0.61 along the full
days axis and 0.33 in
a vertical direction
along the holidays
axis. The unit
(vertex) points are
also shown, as
empty circles on
each axis.*

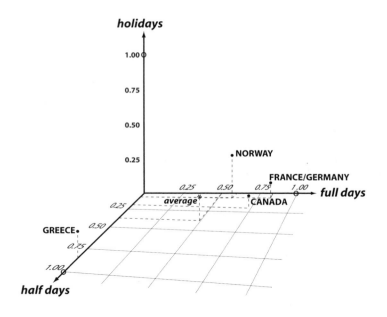

the "wall". All other row profile points in this example, including the average
row profile [0.15 0.36 0.49], can be plotted in this three-dimensional space.

*Vertex points
define the
extremes of the
profile space*
With a bit of imagination it might not be surprising to discover that the
profile points in Exhibit 2.3 all lie exactly in the plane defined by the triangle
that joins the extreme *unit points* [1 0 0], [0 1 0] and [0 0 1] on the three
respective axes, as shown in Exhibit 2.4. This triangle is equilateral and its
three corners are called *vertex points* or *vertices*. The vertices coincide with
extreme profiles that are totally concentrated into one of the day types. For
example, the vertex point [1 0 0] corresponds to a trip to a country consisting
only of *holidays* (fictional in my case, unfortunately). Likewise, the vertex point
[0 0 1] corresponds to a trip consisting only of *full days* of work.

*Triangular (or
ternary)
coordinate system*
Having realized that all profile points in three-dimensional space actually lie
exactly on a flat (two-dimensional) triangle, it is possible to lay this triangle
flat, as in Exhibit 2.5. Looking at the profile points in a flat space is clearly
better than trying to imagine their three-dimensional positions in the corner of
a room! This particular type of display is often referred to as the *triangular* (or
ternary) *coordinate system* and may be used in any situation where we have
sets of data consisting of three elements that add up to 1, as in the case of the
row profiles in this example. Such data are common in geology and chemistry,
for example where samples are decomposed into three constituents, by weight
or by volume. A particular sample is characterized by the three proportions

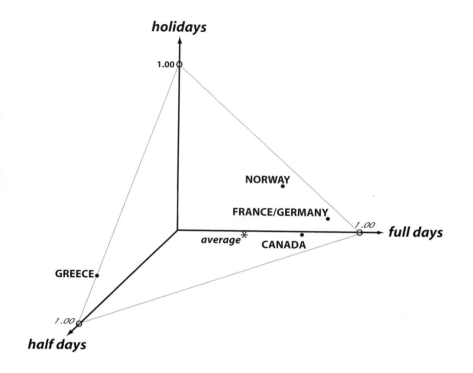

Exhibit 2.4:
The profile points in Exhibit 2.3 lie exactly on an equilateral triangle joining the vertex points of the profile space. Thus the three-dimensional profiles are actually two-dimensional. The profile of Greece lies on the edge of the triangle because it has zero full days.

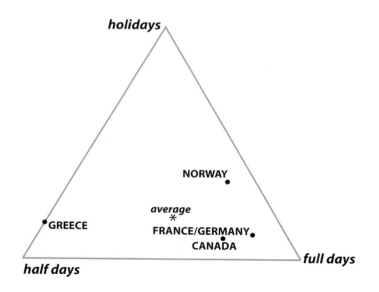

Exhibit 2.5:
The triangle in Exhibit 2.4 that contains the row (country) profiles. The three corners, or vertices, of the triangle represent the columns (day types).

Exhibit 2.6:
*Norway's profile
[0.33 0.06 0.61] is
positioned using
triangular
coordinates as
shown, using the
sides of the triangle
as axes. Each side is
calibrated in profile
units from 0 to 1.*

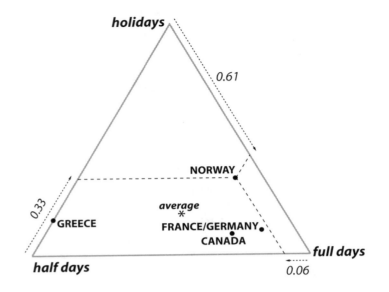

of the constituents and can thus be displayed as a single point with respect to triangular (or ternary) coordinates.

*Positioning a
point in a
triangular
coordinate system*

Given a blank equal-sided triangle and the profile values, how can we find the position of a profile point in the triangle, without passing via the underlying three-dimensional space of Exhibits 2.3 and 2.4? In the triangular coordinate system the sides of the triangle define three axes. Each side is considered to have a length of 1 and can be calibrated accordingly on a linear scale from 0 to 1. In order to position a profile in the triangle, its three values on these axes determine three lines drawn from these values parallel to the respective sides of the triangle. For example, to position Norway, as illustrated in Exhibit 2.5, we take a value of 0.33 on the *holidays* axis, 0.06 on the *half days* axis and 0.61 on the *full days* axis. Lines from these coordinate values drawn parallel to the sides of the triangle all meet at the point representing Norway. In fact, any two of the three profile coordinates are sufficient to situate a profile in this way, and the remaining coordinate is always superfluous, which is another way of demonstrating that the profiles are inherently two-dimensional.

*Geometry of
profiles with more
than three
elements*

The triangular coordinate system may be used only for profiles with three elements. But the idea can easily be generalized to profiles with any number of elements, in which case the coordinate system is known as the *barycentric coordinate system* ("barycentre" is synonymous with "weighted average", to be explained in the next chapter, page 19). The dimensionality of this coordinate system is always one less than the number of elements in the profile. For example, we have just seen that three-element profiles are contained exactly in a two-dimensional triangular profile space. For profiles with four elements the dimensionality is three and the profiles lie in a four-pointed tetrahedron in three-dimensional space. The two-dimensional triangle and the three-dimensional tetrahedron are examples of what is known in mathematics as a *regular simplex*. R code for visualizing an example in three dimensions is given in the Computing Appendix, pages 215–216, so you can get a feeling for three-dimensional profile space. For higher-dimensional profiles some strong imagination would be needed to be able to "see" the profile points spaces of dimension greater than three, but fortunately CA will be of great help to us in visualizing such multidimensional profiles.

Data on a ratio scale

We have illustrated the concept of a profile using frequency data, which is the prime example of data suitable for CA. But CA is applicable to a much wider class of data types; in fact it can be used whenever it makes sense to express the data in relative amounts, i.e., data on a so-called *ratio scale*. For example, suppose we have data on monetary amounts invested by countries in different areas of research — the relative amounts would be of interest, e.g., the percentage invested in environmental research, biomedecine, etc. Another example is of morphometric measurements on a living organism, for example measurements in centimeters on a fish, its length and width, length of fins, etc. Again all these measurements can be expressed relative to the total, where the total is a surrogate measure for the size of the fish, so that we would be analyzing and comparing the shapes of different fish in the form of profiles rather than the original values.

Data on a common scale

A necessary condition of the data for CA is that all observations are on the same scale: for example, counts of particular individuals in a frequency table, a common monetary unit in the table of research investments, centimeters in the morphometric study. It would make no sense in CA to analyze data with mixed scales of measurement, unless a pre-transformation is conducted to homogenize the scales of the whole table. Most of the data sets in this book are frequency data, but in Chapter 23 we shall look at a wide variety of other types of data and ways of recoding them to be suitable for CA.

SUMMARY:
Profiles and the
Profile Space

1. The *profile* of a set of frequencies (or any other amounts that are positive or zero) is the set of frequencies divided by their total, i.e., the set of relative frequencies.

2. In the case of a cross-tabulation, the rows or columns define sets of frequencies which can be expressed relative to their respective totals to give row profiles or column profiles.

3. The marginal frequencies of the cross-tabulation can also be expressed relative to their common total (i.e., the grand total of the table) to give the average row profile and average column profile.

4. Comparing row profiles to their average leads to the same conclusions as comparing column profiles to their average.

5. Profiles consisting of m elements can be plotted as points in an m-dimensional space. Because their m elements add up to 1, these profile points occupy a restricted region of this space. This region is an $(m-1)$-dimensional subspace known as a *simplex*. This simplex is enclosed within the edges joining all pairs of the m unit vectors on the m perpendicular axes. These unit points are also called the *vertices* of the simplex or profile space. The coordinate system within this simplex is known as the *barycentric coordinate system*.

6. A special case that is easy to visualize is when the profiles have three elements, so that the simplex is simply a triangle that joins up three vertices. This special case of the barycentric coordinate system is known as the *triangular* (or *ternary*) *coordinate system*.

7. The idea of a profile can be extended to data on a *ratio scale* where it is of interest to study relative values. In this case the set of numbers being profiled should all have the same scale of measurement.

Masses and Centroids

There is an equivalent way of thinking about the positions of the profile points in the profile space, and this will be useful to our eventual understanding and interpretation of CA. This is based on the notion of a weighted average, or centroid, of a set of points. In the calculation of an ordinary (unweighted) average, each point receives equal weight, whereas a weighted average allows different weights to be associated with each point. When the points are weighted differently, then the centroid does not lie exactly at the "geographical" centre of the cloud of points, but tends to lie in a position closer to the points with higher weight.

Contents

We now use a typical set of data in social science research, a cross-tabulation (or "cross-classification") of two variables from a survey. The table, given in Exhibit 3.1, concerns 312 readers of a certain newspaper, in particular their level of thoroughness in reading the newspaper. Based on data collected in the survey, each respondent was classified into one of three groups: *glance* readers, *fairly thorough* readers and *very thorough* readers. These reading classes have been cross-tabulated against education, an ordinal variable with five categories ranging from some primary education to some tertiary education. Exhibit 3.1 shows the raw frequencies and the education group profiles in parentheses, i.e., the row profiles. The triangular coordinate plot of the row profiles, in the style described in Chapter 2, is given in Exhibit 3.2. In this display the corner points, or vertices, of the triangle represent the three readership groups — remember that each vertex is at the position of a "pure" row profile totally concentrated into that category; for example, the *very thorough* vertex *C3* is representing a fictitious row profile of [0 0 1] that contains 100% *very thorough* readers.

*Data set 2:
Readership and
education groups*

Exhibit 3.1:
*Cross-tabulation of
education group by
readership class,
showing row profiles
and average row
profile in
parentheses, and the
row masses (relative
values of row totals).*

EDUCATION GROUP	Glance C1	Fairly thorough C2	Very thorough C3	Total	Row masses
Some primary E1	5 (0.357)	7 (0.500)	2 (0.143)	14	0.045
Primary completed E2	18 (0.214)	46 (0.548)	20 (0.238)	84	0.269
Some secondary E3	19 (0.218)	29 (0.333)	39 (0.448)	87	0.279
Secondary completed E4	12 (0.119)	40 (0.396)	49 (0.485)	101	0.324
Some tertiary E5	3 (0.115)	7 (0.269)	16 (0.615)	26	0.083
Total	57	129	126	312	
Average row profile	(0.183)	(0.413)	(0.404)		

Exhibit 3.2:
*Row profiles
(education groups)
of Exhibit 3.1
depicted in
triangular
coordinates, also
showing the position
of the average row
profile (last row of
Exhibit 3.1).*

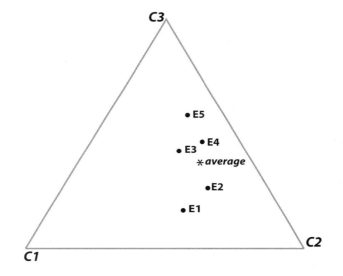

*Points as
weighted averages*

Another way to think of the positions of the education groups in the triangle is as weighted averages. Assigning weights to the values of a variable is a well-known concept in statistics. For example, in a class of 26 students, suppose that the average grade turns out to be 7.5, calculated by summing the 26 grades and dividing by 26. In fact, three students obtain the grade of 9, seven students obtain an 8, and sixteen students obtain a 7, so that the average grade can be determined equivalently by assigning weights of 3/26 to the grade of 9, 7/26 to the grade of 8 and 16/26 to the grade of 7 and then calculating the weighted average. Here the weights are the relative frequencies of each grade, and because the grade of 7 has more weight than the others, the weighted average of 7.5 is "closer" to this grade, whereas the ordinary arithmetic average of the three values 7, 8 and 9 is clearly 8.

Looking at the last row of data in Exhibit 3.1, for education group E5 (some tertiary education), we see the same frequencies of 3, 7 and 16 for the three respective readership groups, and associated relative frequencies of 0.115, 0.269 and 0.615. The idea now is to imagine 3 cases situated at the *glance* vertex *C1* of the triangle, 7 cases at the *fairly thorough* vertex *C2* and 16 cases at the *very thorough* vertex *C3*, and then consider what would be the average *position* for these 26 cases. In other words, we do not associate the weights with values of a variable but with positions in the profile space, in this case the positions of the vertex points. There are more cases at the *very thorough* corner, so we would expect the average position of E5 to be closer to this vertex, as is indeed the case. For the same reason, row profile E1 lies far from the *very thorough* corner *C3* because it has a very low weight (2 out of 14, or 0.143) on this category. Hence each row profile point is positioned within the triangle as an average point, where the profile values, i.e., relative frequencies, serve as the weights allocated to the vertices. Thus, we can think of the profile values not only as coordinates in a multidimensional space, but also as weights assigned to the vertices of a simplex. This idea can be extended to higher-dimensional profiles: for example, a profile with four elements is also at an average position with respect to the four corners of a three-dimensional tetrahedron, weighted by the respective profile elements.

Profile values are weights assigned to the vertices

Alternative terms for weighted average are *centroid* or *barycentre*. Some particular examples of weighted averages in the profile space are given in Exhibit 3.3. For example, the profile point [1/3 1/3 1/3], which gives equal weight to the three corners, is positioned exactly at the centre of the triangle, equidistant from the corners, in other words at the ordinary average position of the three vertices. The profile [1/2 1/2 0] is at a position midway between the first and second vertices, since it has equal weight on these two vertices and zero weight on the third vertex. In general, we can write a verbal formula for the position of a profile as a centroid of the three vertices as follows, for a profile [a b c] where $a + b + c = 1$:

Each profile point is a weighted average, or centroid, of the vertices

$$\text{centroid position} = (a \times \text{vertex 1}) + (b \times \text{vertex 2}) + (c \times \text{vertex 3})$$

For example, the position of education group E5 in Exhibit 3.2 is obtained as follows :

E5 = (0.115 × *glance*) + (0.269 × *fairly thorough*) + (0.615 × *very thorough*)

Similarly, the position of the average profile is also a weighted average of the vertex points:

average = (0.183 × *glance*) + (0.413 × *fairly thorough*) + (0.404 × *very thorough*)

Exhibit 3.3:
*Examples of some
centroids (weighted
averages) of the
vertices in triangular
coordinate space:
the three values are
the weights assigned
to vertices
(C1,C2,C3).*

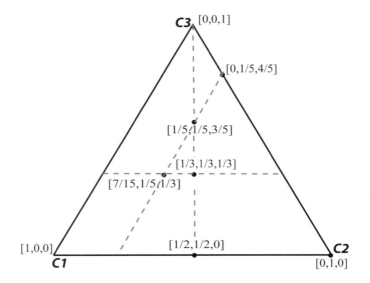

The average is further from the *glance* corner since there is less weight on the *glance* vertex than on the other two, which have approximately the same weights (see Exhibit 3.2).

*Average profile
is also a weighted
average of the
profiles themselves*

The average profile is a rather special point – not only is it a centroid of the three vertices as we have just shown, just like any profile point, but it is also a centroid of the five row profiles themselves, where different weights are assigned to the profiles. Looking again at Exhibit 3.1, we notice that the row totals are different: education group E1 (some primary education) includes only 14 respondents whereas education group E4 (secondary education completed) has 101 respondents. In the last column of Exhibit 3.1, headed "row masses", we have these marginal row frequencies expressed relative to the total sample size 312. Just as we thought of row profiles as weighted averages of the vertices, we can think of each of the five row profile points in Exhibit 3.2 being assigned weights according to their marginal frequencies, as if there were 14 respondents (proportion 0.045 of the sample) at the position E1, 84 respondents (0.269 of the sample) at the position E2, and so on. With these weights assigned to the five profile points, the weighted average position is exactly at the average profile point:

$$Average\ row\ profile = (0.045 \times E1) + (0.269 \times E2) + (0.279 \times E3)$$
$$+ (0.324 \times E4) + (0.083 \times E5)$$

This average row profile is at a central position amongst the row profiles but more attracted to the profiles observed with higher frequency.

*Row and
column masses*

The weights assigned to the profiles are so important in CA that they are given a specific name: *masses*. The last column of Exhibit 3.1 shows the row masses:

0.045, 0.269, 0.279, 0.324 and 0.083. The word "mass" is the preferred term in CA although it is entirely equivalent for our purpose to the term "weight". An alternative term is convenient here to differentiate this geometric concept of weighting from other forms of weighting that occur in practice, such as weights assigned to population subgroups in a sample survey. All that has been said about row profiles and row masses can be repeated in a similar fashion for the columns. Exhibit 3.4 shows the same contingency table as Exhibit 3.1 from the column point of view. That is, the three columns have been expressed in

EDUCATION GROUP	Glance C1	Fairly thorough C2	Very thorough C3	Total	Average column profile
Some primary E1	5 (0.088)	7 (0.054)	2 (0.016)	14	(0.045)
Primary completed E2	18 (0.316)	46 (0.357)	20 (0.159)	84	(0.269)
Some secondary E3	19 (0.333)	29 (0.225)	39 (0.310)	87	(0.279)
Secondary completed E4	12 (0.211)	40 (0.310)	49 (0.389)	101	(0.324)
Some tertiary E5	3 (0.053)	7 (0.054)	16 (0.127)	26	(0.083)
Total	*57*	*129*	*126*	*312*	
Column masses	*0.183*	*0.413*	*0.404*		

Exhibit 3.4:
Cross-tabulation of education group by readership cluster, showing column profiles and average column profile in parentheses, and the column masses

relative frequencies with respect to their column totals, giving three profiles with five values each. The column totals relative to the grand total are now column masses assigned to the column profiles, and the average column profile is the set of row totals divided by the grand total. Again, we could write the average column profile as a weighted average of the three column profiles *C1*, *C2* and *C3*:

$$\textit{Average column profile} = (0.183 \times C1) + (0.413 \times C2) + (0.404 \times C3)$$

Notice how the row and column masses play two different roles, as weights and as averages: in Exhibit 3.4 the average column profile is the set of row masses in Exhibit 3.1, and the column masses here are the elements of what was previously the average row profile.

At this point, even though the final key concepts in CA still remain to be explained, it is possible to make a brief interpretation of Exhibit 3.2. The vertices of the triangle represent the "pure profiles" of readership categories *C1*, *C2* and *C3*, whereas the education groups are "mixtures" of these readership categories and find their positions within the triangle in terms of their respec-

Interpretation in the profile space

tive proportions of each of the three categories. Notice the following aspects of the display:

- The degree of spread of the profile points within the triangle gives an idea of how much variation there is the contingency table. The closer the profile points lie to the centroid, the less variation there is, and the more they deviate from the centroid, the more variation. The profile space is bounded and the most extreme profiles will lie near the sides of the triangle, or in the most extreme case at one of the vertices (for example, an illiterate group would lie on the vertex *C1*). In tables of social science data such as this one, profiles usually occupy a small region of the profile space close to the average because the variation in profile values for a particular category will be relatively small. For example, the range in the first element (i.e., readership category *C1*) across the profiles is only from 0.115 to 0.357 (Exhibit 3.1), in a potential range from 0 to 1. In contrast, for data in ecological research, as we shall see later, the range of profile values is much higher, usually because of many zero frequencies in the table, and the profiles spread out more inside the profile space (see the second example in Chapter 10).

- The profile points are stretched out in what is called a "direction of spread" more or less from the bottom to the top of the display. Looking from the bottom upwards the five education group profiles lie in their natural order (E1 to E5) in order of increasing educational qualifications. At the top, group E5 lies closest to the vertex C3, which represents the highest readership category of *very thorough* reading — we have already seen that this group has the highest proportion (0.615) of these readers. At the bottom, the lower educational group is not far from the edge of the triangle which we know displays profiles with zero *C3* readers (for example, see the point [1/2 1/2 0] in Exhibit 3.3 as an illustration of a point on the edge). The interpretation of this pattern will be that as we move up from the bottom of this display to the top, the profiles are generally changing with respect to their relative frequency of type *C3* as opposed to that of *C1* and *C2* combined, while there is no particular tendency towards either *C1* or *C2*.

Merging rows or columns Suppose we wanted to combine the two categories of primary education, E1 and E2, into a new row of Exhibit 3.1 denoted by E1&2. There are two ways of thinking about this. First, add the two rows together to obtain the row of frequencies [23 53 22] with total 98 and profile [.235 .541 .224]. Second, the profile of E1&2 is the weighted average of of the profiles E1 and E2:

$$[.235 \quad .541 \quad .224] = \frac{.045}{.314} \times [.357 \quad .500 \quad .143] + \frac{.269}{.314} \times [.214 \quad .548 \quad .238]$$

where the masses of E1 and E2 are .045 and .269, with sum .314 (notice that the weights in this weighted average are identical to 14/98 and 84/98, where 14 and 84 are the totals of rows E1 and E2). Geometrically, E1&2's profile lies on a line between E1 and E2, but closer to E2 as shown in Exhibit 3.5.

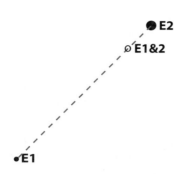

Exhibit 3.5:
*Enlargement of
positions of* E1 *and*
E2 *in Exhibit 3.2,
showing the position
of the point* E1&2
*which merges the
two categories;* E2
*has 6 times the mass
of* E1, *hence* E1&2
lies closer to E2 *at a
point which splits
the line between the
points in the ratio
84:14 = 6:1.*

The distances from E1 to E1&2 and E2 to E1&2 are in the same proportion as the totals 84 and 14, respectively; i.e., 6 to 1. E1&2 can be thought of as the balancing point of the two masses situated at E1 and E2, with the heavier mass at E2.

Suppose that we had an additional row of data in Exhibit 3.1, a category of "no formal education" denoted by E0, with frequencies [10 14 4] across the reading categories. The profile of E0 is identical to E1's profile, because the frequencies in E0 are simply twice those of E1. The two sets of frequencies are said to be *distributionally equivalent.* Thus the profiles of E0 and E1 are at exactly the same point in the profile space, and can be merged into one point with mass equal to the combined masses of the two profiles, i.e., a single point with frequencies [15 21 6].

*Distributionally
equivalent rows or
columns*

The row and column masses are proportional to the marginal sums of the table. If the masses need to be modified for a substantive reason, this can be achieved by a simple transformation of the table. For example, suppose that we require the five education groups of Exhibit 3.1 to have masses proportional to their population sizes rather than their sample sizes. Then the table is rescaled by multiplying each education group profile by its respective population size. The row profiles of this new table are identical to the original row profiles, but the row masses are now proportional to the population sizes. Alternatively, suppose that the education groups are required to be weighted equally, rather than differentially as described up to now. If we regard the table of row profiles (or, equivalently, of row percentages) as the original table, then this table has row sums equal to 1 (or 100%), so that each education group is weighted equally.

*Changing the
masses*

SUMMARY:
Masses and
Centroids

1. We assume that we are concerned with the row problem, i.e., where the row profiles are plotted in the simplex space defined by the column vertices. Then each vertex point represents a column category in the sense that a row profile that is entirely concentrated in that category would lie exactly at that vertex point.

2. Each profile can be interpreted as the *centroid* (or weighted average) of the vertex points, where the weights are the individual elements of the profile. Thus a profile will tend to lie closer to vertices for which it has higher values.

3. Each row profile in turn has a unique weight associated with it, called a *mass*, which is proportional to the row sum in the original table. The average row profile is then the centroid of the row profiles, where each profile is weighted by its mass in the averaging process.

4. Everything described above for row profiles applies equally to the columns of the table. In fact, the best way to make the jump from rows to columns is to rewrite the table in its transposed form, where columns are rows, and vice versa, then everything applies exactly as before.

5. Rows (or columns) that are combined by aggregating their frequencies have a profile equal to the weighted averages of the profiles of the component rows (or columns).

6. Rows (or columns) that have the same profile are said to be *distributionally equivalent* and can be combined into a single point.

7. Row (or column) masses can be modified to be proportional to prescribed values by a simple rescaling of the rows (or columns).

Chi-square Distance and Inertia

In CA the way distance is measured between profiles is a bit more complicated than the one that was used implicitly when we drew and interpreted the profile plots in Chapters 2 and 3. Distance in CA is measured using the so-called *chi-square distance* and this distance is the key to the many favourable properties of CA. There are several ways to justify the chi-square distance: some are more technical and beyond the scope of this book, while other explanations are more intuitive. In this chapter we choose the latter approach, starting with a geometric explanation of the well-known chi-square statistic computed on a contingency table. All the ideas embodied in the chi-square statistic carry over to the chi-square distance in CA as well as to the related concept of inertia, which is the way CA measures variation in a data table.

Contents

Consider the data in Exhibit 3.1 again. Notice that, of the sample of 312 people, 57 (or 18.3%) are in readership category *C1*, 129 (41.3%) in *C2* and 126 (40.4%) in *C3*; i.e., the average row profile is the set of proportions [0.183 0.413 0.404]. If there were no difference between the education groups as far as readership is concerned, we would expect that the profile of each row is more or less the same as the average profile, and would differ from it only because of random sampling fluctuations. Assuming no difference, or in other words assuming that the education groups are *homogeneous* with respect to their reading habits, what would we have expected the frequencies in row E5, for example, to be? There are 26 people in the E5 education group, and we would thus have expected 18.3% of them to be in category *C1*; i.e., $26 \times 0.183 = 4.76$ (although it is ridiculous to talk of 0.76 of a person, it is necessary to maintain such fractions in these calculations). Likewise, we would

Hypothesis of independence or homogeneity for a contingency table

EDUCATION GROUP	Glance C1	Fairly thorough C2	Very thorough C3	Total	Row masses
Some primary E1	5 (2.56)	7 (5.78)	2 (5.66)	14	0.045
Primary completed E2	18 (15.37)	46 (34.69)	20 (33.94)	84	0.269
Some secondary E3	19 (15.92)	29 (35.93)	39 (35.15)	87	0.279
Secondary completed E4	12 (18.48)	40 (41.71)	49 (40.80)	101	0.324
Some tertiary E5	3 (4.76)	7 (10.74)	16 (10.50)	26	0.083
Total	*57*	*129*	*126*	*312*	
Average row profile	*0.183*	*0.413*	*0.404*		

have expected $26 \times 0.413 = 10.74$ of the E5 subjects to be in category *C2*, and $26 \times 0.404 = 10.50$ in category *C3*. There are various names in the literature given to this "assumption of no difference" between the rows of a contingency table (or, similarly, between the columns) — the "hypothesis of independence" is one of them, or perhaps more aptly for our purpose here, the "homogeneity assumption". Under the homogeneity assumption, we would therefore have expected the row of frequencies for E5 to be [4.76 10.74 10.50], but in reality it is observed to be [3 7 16]. In a similar fashion we can compute what each row of frequencies would be if the assumption of homogeneity were exactly true. Exhibit 4.1 shows the expected values in each row underneath their corresponding observed values. Notice that exactly the same expected frequencies are calculated if we argue from the point of view of column profiles, i.e., assuming homogeneity of the readership groups.

*Chi-square
statistic (χ^2) to
test homogeneity
hypothesis*

It is clear that the observed frequencies are always going to be different from the expected frequencies. The question statisticians now ask is whether these differences are large enough to contradict the assumed hypothesis that the rows are homogeneous, in other words whether the discrepancies between observed and expected frequencies are so large that it is unlikely they could have arisen by chance alone. This question is answered by computing a measure of discrepancy between all the observed and expected frequencies, as follows. Each difference between an observed and expected frequency is computed, then this difference is squared and finally divided by the expected frequency. This calculation is repeated for all pairs of observed and expected frequencies and the results are accumulated into a single figure — the *chi-square statistic*, denoted by χ^2:

$$\chi^2 = \sum \frac{(\text{observed} - \text{expected})^2}{\text{expected}}$$

Because there are 15 cells in this 5-by-3 (or 5×3) table, there will be 15 terms in this computation. For purposes of illustration we show only the first three and last three terms corresponding to rows **E1** and **E5**:

$$\chi^2 = \frac{(5 - 2.56)^2}{2.56} + \frac{(7 - 5.78)^2}{5.78} + \frac{(2 - 5.66)^2}{5.66} + \cdots$$

$$+ \frac{(3 - 4.76)^2}{4.76} + \frac{(7 - 10.74)^2}{10.74} + \frac{(16 - 10.50)^2}{10.50} \qquad (4.1)$$

The grand total of the 15 terms in this calculation turns out to be equal to 26.0. The larger this value, the more discrepant the observed and expected frequencies are, i.e., the less convinced we are that the assumption of homogeneity is correct. In order to judge whether this value of 26.0 is large or small we use tables of the chi-square distribution, corresponding to the so-called "degrees of freedom" associated with the statistic. For a 5×3 table, the degrees of freedom are $4 \times 2 = 8$ (one less than the number of rows multiplied by one less than the number of columns), and the P-value associated with the value 26.0 of the χ^2 statistic with 8 degrees of freedom is 0.001. This result tells us that there is an extremely small probability — one in a thousand — that the observed frequencies in Exhibit 4.1 can be reconciled with the homogeneity assumption. In other words, we reject the homogeneity of the table and conclude that it is highly likely that real differences exist between the education groups in terms of their readership profiles.

We are less interested for the moment in the statistical test of homogeneity described above than in the ability of the χ^2 statistic to measure discrepancy from homogeneity, in other words to measure heterogeneity of the profiles. We shall now re-express the χ^2 statistic in a different form by dividing the numerator and denominator of each set of three terms for a particular row by the square of the corresponding row total. For example, looking just at the last three terms of the χ^2 calculation given in (4.1) above, we divide numerator and denominator of each term by the square of **E5**'s total, i.e., 26^2, in order to obtain observed and expected profiles rather than the original raw frequencies:

Alternative expression of the χ^2 statistic in terms of profiles and masses

$$\chi^2 = 12 \text{ similar terms} \cdots + \frac{\left(\frac{3}{26} - \frac{4.76}{26}\right)^2}{\frac{4.76}{26^2}} + \frac{\left(\frac{7}{26} - \frac{10.74}{26}\right)^2}{\frac{10.74}{26^2}} + \frac{\left(\frac{16}{26} - \frac{10.50}{26}\right)^2}{\frac{10.50}{26^2}}$$

$$= 12 \text{ similar terms} \cdots$$

$$+ 26 \times \frac{(0.115 - 0.183)^2}{0.183} + 26 \times \frac{(0.269 - 0.413)^2}{0.413} + 26 \times \frac{(0.615 - 0.404)^2}{0.404}$$

$$(4.2)$$

Notice that one of the factors of 26 in the denominator has been taken out of each of these three terms, leaving all other quantities in the form of profile values. Each of the 15 terms in this calculation is thus of the form:

$$\text{row total} \times \frac{(\text{observed row profile} - \text{expected row profile})^2}{\text{expected row profile}}$$

(Total) inertia is the χ^2 statistic divided by sample size We now make one more modification of the χ^2 calculation above to bring it into line with the CA concepts introduced so far: we divide both sides of the equation (4.2) by the total sample size so that each term involves an initial multiplying factor equal to the row mass rather than the row total:

$$\frac{\chi^2}{312} = 12 \text{ similar terms} \cdots$$

$$+ 0.083 \times \frac{(0.115-0.183)^2}{0.183} + 0.083 \times \frac{(0.269-0.413)^2}{0.413} + 0.083 \times \frac{(0.615-0.404)^2}{0.404}$$
(4.3)

where $0.083 = 26/312$ is the mass of row E5 (see Exhibit 4.1). The quantity χ^2/n on the left-hand side, where n is the grand total of the table, is called the *total inertia* in CA, or simply the *inertia*. It is a measure of how much variance there is in the table and does not depend on the sample size. In statistics this quantity has alternative names such as the mean-square contingency coefficient, and its square root is known as the phi coefficient (ϕ); hence we can denote the inertia by ϕ^2. If we gather together terms in (4.3) in groups of three corresponding to a particular row, we obtain the following form for the inertia:

$$\frac{\chi^2}{312} = \phi^2 = 4 \text{ similar groups of terms} \cdots$$

$$+ 0.083 \times \left[\frac{(0.115 - 0.183)^2}{0.183} + \frac{(0.269 - 0.413)^2}{0.413} + \frac{(0.615 - 0.404)^2}{0.404} \right] (4.4)$$

Each of the five groups of terms in this formula, one for each row of the table, is the row mass (e.g., 0.083 for row E5) multiplied by a quantity in square brackets which looks like a distance measure (or, to be precise, the square of a distance).

Euclidean, or Pythagorian, distance In (4.4) above, if it were not for the fact that each squared difference between observed and expected row profile elements is divided by the expected element, then the quantity in square brackets would be exactly the square of the "straight-line" regular distance between the row profile E5 and the average profile in three-dimensional physical space. This distance is also called the *Euclidean distance* or the *Pythagorian distance*. Let us state this in another way so that it is fully understood. Suppose we plot the two profile points [0.115 0.269 0.615] and [0.183 0.413 0.404] with respect to three perpendicular axes. Then the distance between them would be the square root of the sum of squared differences between the coordinates, as follows:

$$\text{Euclidean distance} = \sqrt{(0.115-0.183)^2 + (0.269-0.413)^2 + (0.615-0.404)^2}$$
(4.5)

This familiar distance, whose value is calculated as 0.264, is exactly the distance between the points E5 and their average in Exhibit 3.2.

However, the distance function in (4.4) is not the Euclidean distance — it involves an extra factor in the denominator of each squared term. Because this factor rescales or *reweights* each squared difference term, this variant of the Euclidean distance function is referred to in general as a *weighted Euclidean distance*. In this particular case where the scaling factors in the denominators are the expected profile elements, the distance is called the *chi-square distance*, or χ^2-distance for short. For example the χ^2-distance between row E5 and the centroid is:

Chi-square distance: an example of a weighted Euclidean distance

$$\chi^2\text{-distance} = \sqrt{\frac{(0.115 - 0.183)^2}{0.183} + \frac{(0.269 - 0.413)^2}{0.413} + \frac{(0.615 - 0.404)^2}{0.404}} \quad (4.6)$$

and has value 0.431, higher than the Euclidean distance in (4.5) because each term under the square root sign has been increased in value. In the next chapter we will show how χ^2-distances can be visualized.

From (4.4) and (4.6) we can write the inertia in the following form:

$$\text{inertia} = \sum_i (i\text{-th mass}) \times (\chi^2\text{-distance from } i\text{-th profile to centroid})^2 \quad (4.7)$$

Geometric interpretation of inertia

where the sum is over the five rows of the table. Since the masses add up to 1, we can also express (4.7) in words by saying that the inertia is the weighted average of the squared χ^2-distances between the row profiles and their average profile. So the inertia will be high when the row profiles have large deviations from their average, and will be low when they are close to the average. Exhibit 4.2 shows a sequence of four small data matrices, each with five rows and three columns, as well as the display of the row profiles in triangular coordinates, going from low to high total inertia. The examples have been chosen especially to illustrate inertias in increasing order of magnitude. This sequence of maps also illustrates the concept of row–column association, or row–column correlation. When the inertia is low, the row profiles are not dispersed very much and lie close to their average profile. In this case we say that there is low association, or correlation, between the rows and columns. The higher the inertia, the more the row profiles lie closer to the column vertices, i.e., the higher is the row–column association. Later, in Chapter 8, we shall describe a correlation coefficient between the rows and columns which links up more formally to the inertia concept.

If all the profiles are identical and thus lie at the same point (their average), all chi-square distances are zero and the total inertia is zero. On the other hand, maximum inertia is attained when all the profiles lie exactly at the vertices of the profile space, in which case the maximum possible inertia is equal to the dimensionality of the space (in the triangular examples of Exhibit 4.2, this maximum would be equal to 2).

Minimum and maximum inertia

Exhibit 4.2:
*A series of data
tables with
increasing total
inertia. The higher
the total inertia, the
greater is the
association between
the rows and
columns, displayed
by the higher
dispersion of the
profile points in the
profile space. The
values in these
tables have been
chosen specifically
so that the column
sums are all equal,
so the weights in the
χ^2-distance
formulation are the
same, and hence
distances we observe
in these maps are
true χ^2-distances.*

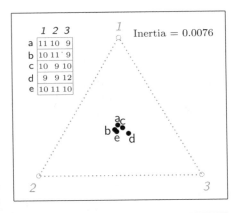

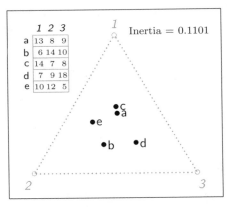

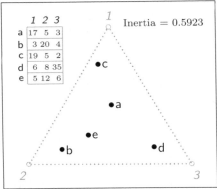

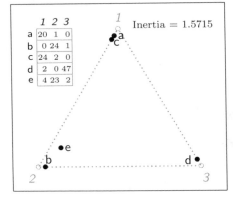

*Inertia of rows
is equal to inertia
of columns*

So far we have explained the concepts of profile, mass, χ^2-distance and inertia in terms of the rows of a data table. As we said in Chapter 3, everything described so far applies in an equivalent way to the columns of the table (see the column profiles, average profile and column masses in Exhibit 3.4). In particular, the calculation of inertia in (4.7) gives an identical result if it is calculated on the column profiles; i.e., the total inertia of a table is the weighted average of the squared χ^2-distance between the column profiles and their average profile, where the weights are now the column masses.

Some notation

This section is not essential to the understanding of the practical aspects of correspondence analysis and may be skipped. But for those who do want to understand the literature on correspondence analysis and its theory, this section will be useful (for example, we shall use these definitions in Chapter 14). We introduce some standard notation for the entities defined so far, using the data in Exhibit 3.1 as illustrations (the data are repeated in Exhibit 4.1).

- n_{ij} — the element of the cross-tabulation (or contingency table) in the i-th row and j-th column, e.g., $n_{21} = 18$.

- n_{i+} — the total of the i-th row, e.g., $n_{3+} = 87$ (the $+$ in the subscript indicates summation over the corresponding index).

- n_{+j} — the total of the j-th column, e.g., $n_{+2} = 129$.

- n_{++}, or simply n, the grand total of the table, e.g., $n = 312$.

- p_{ij} — the n_{ij} divided by the grand total of the table, e.g., $p_{21} = n_{21}/n = 18/312 = 0.0577$.

- r_i — the mass of the i-th row, i.e., $r_i = n_{i+}/n$ (which is the same as p_{i+}, the sum of the i-th row of relative frequencies p_{ij}); e.g., $r_3 = 87/312 = 0.279$; the vector of row masses is denoted by $\mathbf{r}$.

- c_j — the mass of the j-th column, i.e., $c_j = n_{+j}/n$ (which is the same as p_{+j}, the sum of the j-th column of relative frequencies p_{ij}); e.g., $c_2 = 129/312 = 0.414$; the vector of column masses is denoted by $\mathbf{c}$.

- a_{ij} — the j-th element of the profile of row i, i.e., $a_{ij} = n_{ij}/n_{i+}$; e.g., $a_{21} = 18/84 = 0.214$; the profile is denoted by the vector $\mathbf{a}_i$.

- b_{ij} — the i-th element of the profile of column j, i.e., $b_{ij} = n_{ij}/n_{+j}$; e.g., $b_{21} = 18/57 = 0.316$; the profile is denoted by the vector $\mathbf{b}_j$.

- $\sqrt{\sum_j (a_{ij} - a_{i'j})^2/c_j}$ — the χ^2-distance between the i-th and i'-th row profiles, denoted by $\|\mathbf{a}_i - \mathbf{a}_{i'}\|_c$; e.g., from Exhibit 3.1

$$\|\mathbf{a}_1 - \mathbf{a}_2\|_c = \sqrt{\frac{(0.357-0.214)^2}{0.183} + \frac{(0.500-0.548)^2}{0.413} + \frac{(0.143-0.238)^2}{0.404}} = 0.374.$$

- $\sqrt{\sum_i (b_{ij} - b_{ij'})^2/r_i}$ — the χ^2-distance between the j-th and j'-th column profiles, denoted by $\|\mathbf{b}_j - \mathbf{b}_{j'}\|_r$; e.g., from Exhibit 3.4

$$\|\mathbf{b}_1 - \mathbf{b}_2\|_r = \sqrt{\frac{(0.088-0.054)^2}{0.045} + \frac{(0.316-0.357)^2}{0.269} + \ldots \text{etc.}} = 0.323$$

where $0.088 = 5/57$, $0.054 = 7/129$, $0.045 = 14/312$, etc.

- $\sqrt{\sum_j (a_{ij} - c_j)^2/c_j}$ — the χ^2-distance between the i-th row profile $\mathbf{a}_i$ and the average row profile $\mathbf{c}$ (the vector of column masses), denoted by $\|\mathbf{a}_i - \mathbf{c}\|_c$; e.g., from Exhibit 3.1

$$\|\mathbf{a}_1 - \mathbf{c}\|_c = \sqrt{\frac{(0.357-0.183)^2}{0.183} + \frac{(0.500-0.413)^2}{0.413} + \frac{(0.143-0.404)^2}{0.404}} = 0.594.$$

- $\sqrt{\sum_i (b_{ij} - r_i)^2/r_i}$ — the χ^2-distance between the j-th column profile $\mathbf{b}_j$ and the average column profile $\mathbf{r}$ (the vector of row masses), denoted by $\|\mathbf{b}_j - \mathbf{r}\|_r$; e.g., from Exhibit 3.4

$$\|\mathbf{b}_1 - \mathbf{r}\|_r = \sqrt{\frac{(0.088-0.045)^2}{0.045} + \frac{(0.316-0.269)^2}{0.269} + \ldots \text{etc.}} = 0.332.$$

With this notation, the formula (4.7) for the total inertia is

$$\phi^2 = \frac{\chi^2}{n} = \sum_i r_i \|\mathbf{a}_i - \mathbf{c}\|_c^2 \qquad \text{(for the rows)}$$

$$= \sum_i r_i \sum_j \left(\frac{p_{ij}}{r_i} - c_j \right)^2 / c_j \qquad (4.8)$$

$$= \sum_j c_j \|\mathbf{b}_i - \mathbf{r}\|_r^2 \qquad \text{(for the columns)}$$

$$= \sum_j c_j \sum_i \left(\frac{p_{ij}}{c_j} - r_i \right)^2 / r_i \qquad (4.9)$$

and has the value 0.0833, hence $\chi^2 = 0.0833 \times 312 = 26.0$.

SUMMARY:
Chi-square
Distance and
Inertia

1. The chi-square (χ^2) statistic is an overall measure of the difference between the observed frequencies in a contingency table and the expected frequencies calculated under a hypothesis of homogeneity of the row profiles (or of the column profiles).

2. The *(total) inertia* of a contingency table is the χ^2 statistic divided by the total of the table.

3. Geometrically, the inertia measures how "far" the row profiles (or the column profiles) are from their average profile. The average profile can be considered to represent the hypothesis of homogeneity (i.e., equality) of profiles.

4. Distances between profiles are measured using the *chi-square distance* (χ^2-distance). This distance is similar in formulation to the *Euclidean* (or *Pythagorian*) distance between points in physical space, except that each squared difference between coordinates is divided by the corresponding element of the average profile.

5. The inertia can be rewritten in a form which can be interpreted as the weighted average of squared χ^2-distances between the row profiles and their average profile (similarly, between the column profiles and their average).

Plotting Chi-square Distances

In Chapter 3 we interpreted the positions of two-dimensional profile points in a triangular coordinate system where distances were Euclidean distances. In Chapter 4 the chi-square distance (χ^2-distance) between profile points was defined, as well as its connection with the chi-square statistic and the inertia of a data matrix. The χ^2-distance is a weighted Euclidean distance, where each squared term corresponding to a coordinate is weighted inversely by the average profile value corresponding to that coordinate. So far we have not actually visualized the χ^2-distances between profiles, apart from Exhibit 4.2, where the average profile values were equal, so that the χ^2-distances were also Euclidean in that case. In this chapter we show that by a simple transformation of the profile space, the distances that we observe in our graphical display are actual χ^2-distances.

Contents

Exhibit 5.1 shows the row profiles of Exhibit 3.1 plotted according to perpendicular coordinate axes in the usual three-dimensional physical space. Here the distances between the profiles are not χ^2-distances, but rather (unweighted) Euclidean distances — see the calculation in formula (4.5). In such a space distances between two profiles with elements x_j and y_j respectively (where $j = 1, \ldots, J$) are calculated by summing the squared differences between coordinates, of the form $(x_j - y_j)^2$, over all dimensions j and then taking the square root of the resultant sum. This is the usual "straight-line" Euclidean distance of physical space with which we are familiar. As we have seen, the χ^2-distance differs from this distance function by the division of each squared difference by the corresponding element of the average profile; i.e., each term is of the form $(x_j - y_j)^2/c_j$, where c_j is the corresponding element of the average profile. Since we can interpret and compare distances only in our familiar physical space, we need to be able to organize the points in the map

Difference between χ^2-distance and ordinary Euclidean distance

Exhibit 5.1:
*The profile space
showing the profiles
of the education
groups on the
equilateral triangle
in three-dimensional
space; the distances
here are Euclidean
distances.*

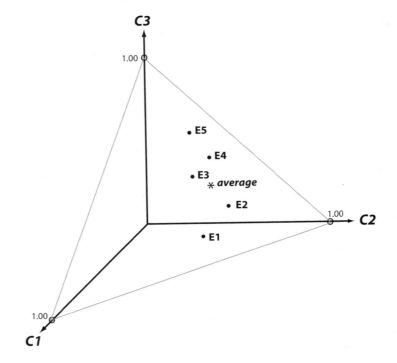

Exhibit 5.1:
The profile space showing the profiles of the education groups on the equilateral triangle in three-dimensional space; the distances here are Euclidean distances.

in such a way that familiar "straight-line" distances turn out to be the χ^2-distances. Luckily, this is possible thanks to a straightforward transformation of the profiles.

Transforming the coordinates before plotting — In the calculation of the χ^2-distance every term of the form $(x_j - y_j)^2/c_j$ can be rewritten as $(x_j/\sqrt{c_j} - y_j/\sqrt{c_j})^2$. This equivalent way of expressing the general term in the distance formula is identical in form to that of the ordinary Euclidean distance function; i.e., it is in the form of a squared difference. The only change is that the coordinates are not the original x_j and y_j values but the transformed $x_j/\sqrt{c_j}$ and $y_j/\sqrt{c_j}$ ones. This suggests that, instead of using the elements of the profiles as coordinates, we should rather use these elements divided by the square roots of the corresponding elements of the average profile. In that case the usual Euclidean distance between these transformed coordinates gives the χ^2-distance that we require.

Effect of the transformation in practice — The values of c_j are elements of the average profile and thus all less than 1. So the transformation of the profile elements by dividing by $\sqrt{c_j}$ will result in an increase in the values of all coordinates, but some will be increased more than others. If a particular c_j is relatively small compared to the others (i.e., the j-th column category has a relatively low frequency), then the corresponding coordinates $x_j/\sqrt{c_j}$ and $y_j/\sqrt{c_j}$ will be increased by a relatively large amount. Conversely, a large c_j corresponding to a more frequent category will lead to a relatively smaller increase in the transformed coordinates. Thus the effect of the transformation is to increase the values corresponding to low-frequency categories relatively more than the coordinates corresponding to high-frequency categories. In the untransformed space of Exhibit 5.1 the vertex points lie at one physical unit of measurement from the origin (i.e., zero point) of the three coordinates axes. The first vertex, with coordinates $[1 \; 0 \; 0]$, is transformed to the position $[1/\sqrt{c_1} \; 0 \; 0]$; i.e., its position on the first axis is stretched out to be $1/\sqrt{0.183} = 2.34$. Similarly, the second and third vertices are stretched out to values $1/\sqrt{c_2} = 1/\sqrt{0.413} = 1.56$ and $1/\sqrt{c_3} = 1/\sqrt{0.404} = 1.57$, respectively. These values are indicated at the vertices on the three axes in Exhibit 5.2. The profiles are plotted according to their transformed values and find new positions in the space, but are still in the triangle joining the transformed vertex points. Notice that the stretching is relatively more in the direction of $C1$, the category with the lowest marginal frequency.

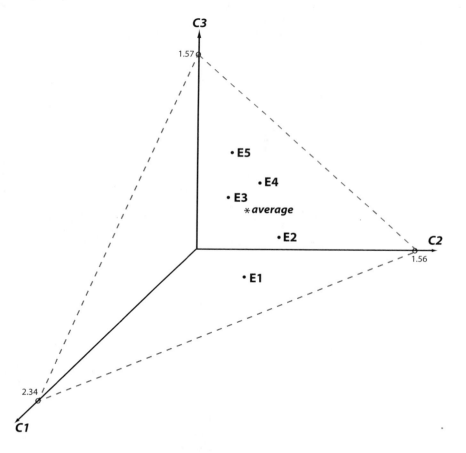

Exhibit 5.2:
The profile space showing the axes stretched by different amounts so that distances between profiles are χ^2-distances.

There is an equivalent way of thinking of this situation geometrically. In the untransformed coordinate systems of Exhibits 2.4 and 5.1, the tic marks indicating the scales (for example, the values 0.1, 0.2, 0.3, etc.) along the three axes were at equal intervals apart. The effect of the transformation is to stretch out the three vertices as shown in Exhibit 5.2. But we can still think of the three vertex points as being one profile unit from the origin, but then the scales are different on the three axes. On the *C1* axis, an interval of 0.1 between two tic marks would be a physical length of 0.234, while on the *C2* and *C3* axes these intervals would be 0.156 and 0.157, respectively. Hence a unit interval on the *C1* axis is approximately 50% longer than the same interval on the other two axes. Along these recalibrated axes we would still use the original profile elements to situate a profile in three-dimensional space. Whichever way you prefer to think about the transformation, either as a transformation of the profile values or as a stretching and recalibrating of the axes, the outcome is the same: the profile points now lie in the stretched triangular space shown in Exhibit 5.2. In Exhibit 5.3 the stretched triangle has been laid flat and it is clear that vertex *C1*, corresponding to the rarest category of *glance* reading, has been stretched the most.

Exhibit 5.3:
*The triangular space
of the profiles in the
stretched space of
Exhibit 5.2 laid
"flat" (compare with
Exhibit 3.2). The
triangle has been
stretched most in
the direction of C1,
the category with
the lowest frequency.*

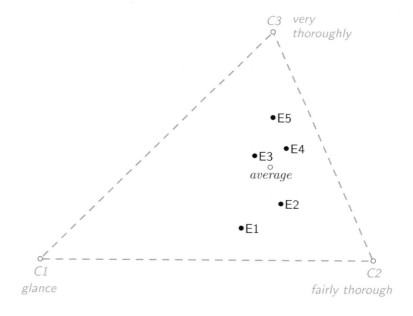

Now that the observed straight-line distances in the transformed space are actual χ^2-distances, the profile points may be joined to the average point to show the χ^2-distances between the profiles and their average — see Exhibit 5.4. In the case of row profiles, if we associate each row mass with its respective profile, we know from formula (4.7) that the weighted sum of these squared distances is identical to the inertia of the table. If we associate the total row frequencies with the profiles rather than the masses (where the total row frequency is n times the row mass, n being the grand total of the whole table), then the weighted sum of these squared distances is equal to the χ^2 statistic. Equivalent results hold for the column profiles relative to their average point. Thus the inertia and χ^2 statistic may be interpreted geometrically as the degree of dispersion of the set of profile points (rows or columns) about their average, where the points are weighted proportional to their relative frequency.

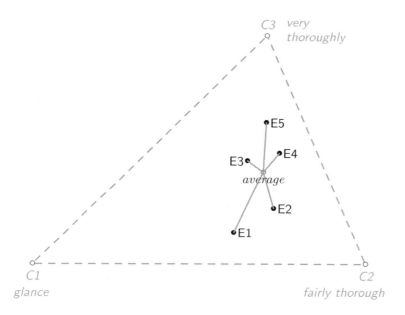

Exhibit 5.4:
The "stretched" profile space showing the χ^2-distances from the profiles to their centroid; the inertia is the weighted average of the sum of squares of these distances, and the χ^2 statistic is the inertia multiplied by the sample size ($n = 312$ in this example).

To explain this principle, consider Exhibit 3.1 again and suppose that we could distinguish two types of *fairly thorough* readers, those that concentrated more on the political content of the newspaper and those that concentrated more on the cultural and sports sections; these categories are denoted by *C2a* and *C2b*, respectively. Suppose further that in both these new columns, the relative frequencies of education groups were the same; in other words, suppose that there was no difference between these two subdivisions of the *fairly thorough* reading group as far as education is concerned. In Chapter 3 we called such columns distributionally equivalent, since they have the same profiles. The subdivision of column *C2* into *C2a* and *C2b* brings no extra information about the differences between the education groups; hence any analysis of these data should give the same results whether *C2* is subdivided or left as a single category. An analysis that satisfies this property is said to obey the *principle of distributional equivalence*. If we used ordinary Euclidean distances between the education group profiles, this principle would not be

Principle of distributional equivalence

obeyed because different results would be obtained if such a subdivision were made. The χ^2-distance, on the other hand, does obey the principle, remaining unaffected by such a partitioning of a category of the data matrix: if two distributionally equivalent columns are merged, the χ^2-distance between the rows does not change. In practice, this means that columns that have similar profiles can be aggregated without affecting the geometry of the rows, and vice versa. This gives the researcher a certain assurance that introducing many categories into the analysis adds only substantive value and is not affected by some technical quirk that depends on the number of categories.

χ^2-distances make the contributions of categories more similar

We know now how to organize the display in order to observe χ^2-distances, but why do we need to go to all this trouble to visualize χ^2-distances rather than Euclidean distances? There are many ways to defend the use of the χ^2-distance, some more technical than others, and the reason is more profound than simply being able to visualize the χ^2 statistic. There are inherent disparities in the variances of sets of frequencies. For example, in Exhibit 3.1 one can see that the range of profile values in the less frequent column *C1* (from 0.115 to 0.357) is less than the range in the more frequent column *C3* (from 0.143 to 0.615). This is a general rule for frequency data, namely that a set of smaller frequencies has less dispersion than a set of larger frequencies. The effect of this disparity on the spread of the profile values can be seen by measuring the contributions of each category to the distance function. For example, let us compare the squared values of the Euclidean distances and the χ^2-distances between the education group profiles and their centroid (average profile) in the data set of Exhibit 3.1. For example, for the fifth education group *E5*, the squared Euclidean distance between its profile and the centroid is:

$$
\begin{aligned}
\text{Euclidean distance}^2 &= (0.115 - 0.183)^2 + (0.269 - 0.413)^2 + (0.615 - 0.404)^2 \\
&= 0.00453 + 0.02080 + 0.04475 \\
&= 0.07008
\end{aligned}
$$

while the squared χ^2-distance is:

$$
\begin{aligned}
\chi^2\text{-distance}^2 &= \frac{(0.115 - 0.183)^2}{0.183} + \frac{(0.269 - 0.413)^2}{0.413} + \frac{(0.615 - 0.404)^2}{0.404} \\
&= 0.02480 + 0.05031 + 0.11081 \\
&= 0.18592
\end{aligned}
$$

(see (4.5) and (4.6) on pages 28–29). Each of these squared distances is the sum of three values, one term for each column category, and can be expressed as percentages of the total to assess the contributions of each category of readership. For example, in the squared Euclidean distance category *C1* contributes 0.00453 out of 0.07008, which is 6.5%, whereas in the squared χ^2-distance its contribution is 0.02480 out of 0.18592, i.e., 13.3% (see the row *E5* in Exhibit 5.5). If all the terms for *C1* are summed over the five education groups and expressed as a percentage of the sum of squared distances we get the overall percentage contribution of 17.0% for the Euclidean distance, and 31.3% for the χ^2-distance, given in the last row of Exhibit 5.5. This exercise illustrates the phenomenon that the lowest frequency category *C1* generally contributes less to the Euclidean distance compared to *C3*, for example, whereas in the χ^2-distance its contribution is boosted owing to the division by the average frequency.

Row	Euclidean			χ^2		
	C1	*C2*	*C3*	*C1*	*C2*	*C3*
E1	28.7	7.1	64.2	47.1	5.1	47.7
E2	2.1	38.7	59.1	4.7	37.2	58.1
E3	13.2	66.4	20.4	25.5	56.7	17.8
E4	37.1	2.8	60.1	56.6	1.9	41.5
E5	6.5	29.7	63.9	13.3	27.1	59.6
Overall	17.0	21.8	61.2	31.3	17.7	51.0

Exhibit 5.5: *Percentage contributions of each column category to the squared Euclidean and squared χ^2-distances from the row profiles to their centroid (data of Exhibit 3.1).*

As described in Chapter 4, the χ^2-distance is an example of a weighted Euclidean distance, whose general definition is as follows:

Weighted Euclidean distance

$$\text{weighted Euclidean distance} = \sqrt{\sum_{j=1}^{p} w_j (x_j - y_j)^2} \qquad (5.1)$$

where w_j are nonnegative weights and x_j, $j = 1, \ldots, p$ and y_j, $j = 1, \ldots, p$ are two points in p-dimensional space. In principal component analysis (PCA), a method closely related to CA, the p dimensions are defined by continuous variables, often on different measurement scales. It is necessary to remove the effect of scale in some way, and this is usually done by dividing the data by the standard deviations s_j of the respective variables; i.e., observations x_j and y_j for variable j are replaced by x_j/s_j and y_j/s_j. This operation can be thought of as using a weighted Euclidean distance with weights $w_j = 1/s_j^2$, the inverse of the variances. In the definition of the χ^2-distance between profiles, the weights are equal to $w_j = 1/c_j$, i.e., the inverse of the mean profile elements.

Although the profiles are on the same relative frequency scale, there is still a need to compensate for different variances, similar to the situation in PCA. The phenomenon that sets of frequencies with higher average have higher variance than those with a lower average is embodied in the *Poisson distribution* — one of the standard statistical distributions for variables that are counts. A property of the Poisson distribution is that its variance is equal to its mean. Hence transforming the frequencies by dividing by the square roots of the expected (mean) frequencies is one way of standardizing the data because the square root of the mean is a surrogate for the standard deviation. But it is not the only way to standardize, so why is the χ^2-distance so special? There are

Theoretical justification of χ^2-distance

many advantages of the χ^2-distance, apart from its obeying the principle of distributional equivalence and giving CA the property of symmetry between the treatment of rows and columns. A more technical reason for using the χ^2-distance can be found in the properties of a multivariate statistical distribution for count data, called the *multinomial distribution*. This subject will be discussed in more detail in the Theoretical Appendix.

SUMMARY:
Plotting
Chi-square
Distances

1. χ^2-distances between profiles can be observed in ordinary physical (or Euclidean) space by transforming the profiles before plotting. This transformation consists of dividing each element of the profile by the square root of the corresponding element of the average profile.

2. Another way of thinking about χ^2-distances is not to transform the profile elements but to stretch the plotting axes by different amounts, so that a unit on each axis has a physical length inversely proportional to the square root of the corresponding element of the average profile.

3. The χ^2-distance is a special case of a weighted Euclidean distance where the weights are the inverses of the corresponding average profile values.

4. Assuming that we are plotting row profiles, the rescaling of the coordinates (or, equivalently, the stretching of the axes) can be regarded as a way of standardizing the columns of the table. This makes visual comparisons between the row profiles more equitable across the different columns.

5. The χ^2-distance obeys the *principle of distributional equivalence*, which guarantees stability in the distances between rows, say, when columns are partitioned into similar components, or when similar columns are merged.

Reduction of Dimensionality

Up to now, small data sets (Exhibits 2.1 and 3.1) were used specifically because they were low-dimensional and hence easy to visualize exactly. These tables with three columns involve three-dimensional profiles, which are actually two-dimensional, as we saw in Chapter 2, and can thus be laid flat for inspection in a triangular coordinate system. In most applications, however, the table of interest has many more rows and columns and the profiles lie in a space of much higher dimensionality. Since we cannot easily observe or even imagine points in a space with more than three dimensions, it becomes necessary to reduce the dimensionality of the points. This dimension-reducing step is the crucial analytical aspect of CA and can be performed only with a certain loss of information, but the objective is to restrict this loss to a minimum so that a maximum amount of information is retained.

Contents

An example of a table of higher dimensionality is given in Exhibit 6.1, a cross-tabulation generated from the database of the Spanish National Health Survey (*Encuesta Nacional de la Salud*) in 1997. One of the questions in this survey concerns the opinion that respondents have of their own health, which they can judge to be "very good" (*muy bueno* in the original survey), "good" (*bueno*), "regular" (*regular*), "bad" (*malo*) or "very bad" (*muy malo*). The table cross-tabulates these responses with the age groups of the respondents. There are seven age groups (rows of Exhibit 6.1) and five health categories (columns). A total of 6371 respondents are cross-tabulated and give a representative snapshot of how the Spanish nation views its own health at this point in time. But what is that view, and how does it change with age? Using CA

Data set 3: Spanish National Health Survey

Exhibit 6.1:
*Cross-tabulation of
age group with
self-perceived health
category. Data
source: Spanish
National Health
Survey (Encuesta
Nacional de la
Salud), 1997.*

AGE GROUP	Very good	Good	Regular	Bad	Very bad	Sum
16–24	243	789	167	18	6	*1223*
25–34	220	809	164	35	6	*1234*
35–44	147	658	181	41	8	*1035*
45–54	90	469	236	50	16	*861*
55–64	53	414	306	106	30	*909*
65–74	44	267	284	98	20	*713*
75+	20	136	157	66	17	*396*
Sum	*817*	*3542*	*1495*	*414*	*103*	*6371*

Exhibit 6.2:
*Profiles of age
groups across the
health categories,
expressed as
percentages.*

AGE GROUP	Very good	Good	Regular	Bad	Very bad	Sum
16–24	19.9	64.5	13.7	1.5	0.5	*100.0*
25–34	17.8	65.6	13.3	2.8	0.5	*100.0*
35–44	14.2	63.6	17.5	4.0	0.8	*100.0*
45–54	10.5	54.5	27.4	5.8	1.9	*100.0*
55–64	5.8	45.5	33.7	11.7	3.3	*100.0*
65–74	6.2	37.4	39.8	13.7	2.8	*100.0*
75+	5.1	34.3	39.6	16.7	4.3	*100.0*
Average	*12.8*	*55.6*	*23.5*	*6.5*	*1.6*	*100.0*

we will be able to understand very quickly the relationship between age and self-perception of health.

*Comparison of
age group (row)
profiles*

Let us suppose for the moment that we are interested in the profiles of the age groups across the health categories, i.e., the row profiles. The row profiles are given in percentage form in Exhibit 6.2. The last row is the average row profile, or the profile across the health categories for the sample as a whole, without distinguishing between age groups. Thus we can see, for example, that of the total of 6371 Spaniards sampled in this study, 12.8% regarded themselves as in *very good* health, 55.6% in *good* health, and so on. Looking at specific age groups we see that there are the differences that one would expect; for example, the youngest age group has higher percentages of these categories (19.9% *very good* and 64.5% *good*) whereas the oldest group has lower percentages (5.1% and 34.3%, respectively). Perusing this table we quickly come to the conclusion that self-perceived health becomes worse with age, which is no surprise at all. It is not so easy, however, to see in the numbers how fast or slow this change is occurring; for example, where the changes in self-perceived health from one age group to the next are bigger or smaller.

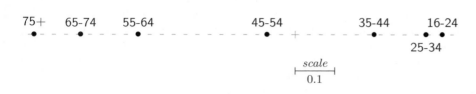

CA visualizes the age groups and gives us more insight into the data. The problem in this example is that one cannot visualize the age group profiles exactly, since they are points situated in a five-dimensional space. Actually, as we saw in the previous three-dimensional examples, the five-element age group profiles lie in a space of one less dimension because the elements of each profile add to 1, but even direct visualization in four-dimensional space is impossible. We might be able to visualize the profiles approximately, however, hoping that they do not "fill" the whole four-dimensional space but rather lie approximately in some low-dimensional subspace of one, two or three dimensions. This is the essence of CA, the identification of a low-dimensional subspace which approximately contains the profiles. Putting this the opposite way, CA identifies dimensions along which there is very little dispersion of the profile points and eliminates these low-information directions of spread, thereby reducing the dimensionality of the cloud of points so that we can more easily visualize their relative positions.

*Identifying
lower-dimensional
subspaces*

In this example it turns out that the profiles actually lie very close to a line, so that the points can be imagined as forming an elongated cigar-shaped cloud of points in the four-dimensional profile space. If we now identify the line which comes "closest" to the points (we define the measure of closeness soon), we can drop (or *project*) the points perpendicularly onto this line, take the line out of the multidimensional space and lay it from left to right on a display which is now much easier to interpret. In Exhibit 6.3 we see this one-dimensional representation of the age group profiles, with the age groups lying in their inherent order from oldest on the left to youngest on the right, even though the method has no knowledge of the ordering of the categories. In this display we can see immediately that there are smaller differences amongst the younger age groups, and bigger differences in the middle-age groups.

*Projecting profiles
onto subspaces*

Since the lower-dimensional projections of the profiles are no longer at their true positions, we need to know how large a discrepancy there is between their exact positions and their approximate ones. To do this we use the total inertia of the profiles as a measure of the total variation, or geometric dispersion, of the points in their true four-dimensional positions. Both quality of display and its counterpart, the loss, or error of display, are measured in the form of percentages of the total inertia, and they add up to 100%: the lower the loss,

*Measuring quality
of display*

the higher the quality, and the higher the loss the lower the quality. In the present example the loss incurred by projecting the points onto the straight line of Exhibit 6.3 turns out to be only 2.7%; in other words the quality of the unidimensional approximation of the profiles is equal to 97.3%. This is a very favourable result — we started with a 7×5 table of numbers with an inherent dimensionality of 4 and, by sacrificing only 2.7% of the dispersion of the points in three dimensions of the space, the remaining 97.3% is represented by a scatter of points along a single dimension! This percentage can be interpreted exactly as in regression as a "percentage of explained variance": the single dimension showing the seven projected profile points in Exhibit 6.3 explains 97.3% of the inertia of the true profiles (or 97.3% of the total inertia of the table in Exhibit 6.1).

Exhibit 6.4:
Observed interpoint distances measured in Exhibit 6.3 between all pairs of points, plotted against the true χ^2-distances between the row profiles of Exhibit 6.3.

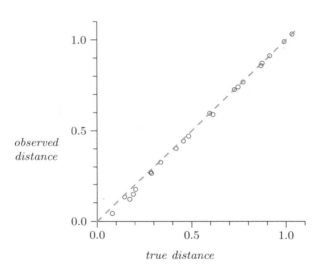

Approximation of interprofile distances

The distances between the projected profiles in Exhibit 6.3 are approximations of the true χ^2-distances between the row profiles in their full four-dimensional space. Exact χ^2-distances, computed directly from Exhibit 6.2, can be compared with those displayed in Exhibit 6.3, and this comparison is made graphically in Exhibit 6.4. Because there are 7 points there are $\frac{1}{2} \times 7 \times 6 = 21$ pairs of interpoint distances. Clearly the agreement is excellent, which was expected because of the relatively small loss in accuracy of 2.7% incurred in reducing the profiles to a one-dimensional display. Notice in Exhibit 6.4 that the observed distances are always less than or equal to the true distances — we say that the distances are approximated "from below". This is because the square of the true distance is the sum of a set of squared components, one for each dimension of the profile space, whereas the square of the observed distance is the sum of a reduced number of these components, which in this unidimensional example is just a single component. The "unexplained" part of the distances is shown by the deviations of the points from the 45° line in Exhibit 6.4.

In the space of the seven age group profiles, there are five vertex points representing the health categories. Recall once more that each of these extreme profile points represents a fictitious profile totally concentrated into one health category; for example, the vertex point [1 0 0 0 0] represents a group which has only *very good* self-perceived health. These vertex points can also be projected onto the dimension in Exhibit 6.3 which best explains the age group profiles — see Exhibit 6.5. Notice the change in scale compared to Exhibit 6.3 — the age group profiles are in exactly the same positions in both these maps. The vertices are much more spread out than the profiles because they are the most extreme profiles obtainable.

Display of the projected vertex points

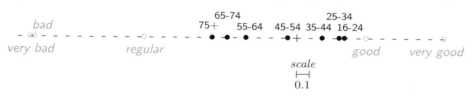

Exhibit 6.5:
Optimal map of Exhibit 6.3, showing the projected vertices of the health categories.

Notice how in the joint display of Exhibit 6.5 the health categories are also spread out in their inherent order, with the *very bad* health category on the extreme left and the *very good* on the extreme right. The positions of these reference points along the dimension gives us the key to the interpretation of the association between the rows (age groups) and columns (health categories), with the youngest age group furthest towards good health and the oldest group furthest towards bad health. The origin (or zero point, indicated by a + in Exhibits 6.3 and 6.5) represents the average profile; thus we can deduce that the age groups up to 44 years are on the "good" side of average, and those from 45 years up on the "bad" side. The fact that *very bad* is so far away from the age group profiles shows that no age group is close to this extreme — indeed in Exhibit 6.2 we can see percentage values of 0.5–4.3% and an average of 1.6% (the average value is at the origin). The category *bad* is almost at the same position, but with a range of 1.5–16.7% and an average of 6.5% at the origin (more details about the joint interpretation will be given in Chapters 8 and 13). The relationship between the row profiles and column vertices in this one-dimensional projection is the same as we described for the triangular space in Chapters 2 and 3 — each age group profile is at the weighted average of the health category vertices, using the profile elements as weights. Hence the youngest age group **16–24** is at the rightmost position of the age groups

Joint interpretation of profiles and vertices

Exhibit 6.6:
*Profile points in a
multidimensional
space and a plane
cutting through the
space; the
best-fitting plane in
the sense of least
squares must pass
through the centroid
of the points.*

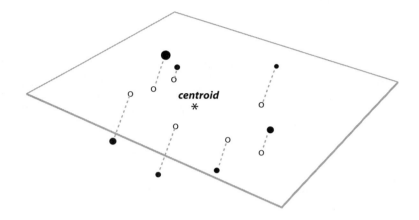

because it has the highest profile values on the health categories *very good*
and *good* on the right.

*Definition of
closeness of points
to a subspace*

The present example is simpler than usual because a single dimension ade-
quately summarizes the data. In most cases we shall look for at least a two-
dimensional plane that comes "closest to", or "best fits", the high-dimensional
cloud of profiles. The profiles are then projected onto this plane, and the extreme vertices of the profile space as well. Exhibit 6.6 shows several profile
points in an imaginary high-dimensional space, and their projections onto a
plane cutting through the space. Whether we project the profiles onto a best-
fitting line (a one-dimensional subspace), a plane (two-dimensional subspace)
or even a subspace of higher dimensionality, we need to define what we mean
by "closeness" of the points to that subspace. Imagine any straight line in
the multidimensional space of the profiles. The shortest distance from each
profile to the line can be computed (by distance in this context we implicitly
mean the χ^2-distance). To arrive at a single measure of closeness of all points
to the line, an intuitively obvious choice would be to add the distances from
all profiles to the imaginary line. Then our task would be to find the line
for which this sum-of-distances is the smallest. In principle there is nothing
stopping us from doing exactly this, but the mathematics involved in minimizing such a sum-of-distances is quite complicated. As in many other areas
of statistics the problem simplifies greatly if one defines a criterion in terms
of sum of squared distances, rather than the distances alone, leading to what
is called a *least-squares* problem. In the present case, we also have a mass
associated with each profile which quantifies the importance of the profile in
the analysis. The criterion used in correspondence analysis is thus a weighted
sum of squared distances criterion.

Suppose that we have I profile points in a multidimensional space and that a candidate low-dimensional subspace is denoted by S. For the i-th profile point, with mass m_i, we compute the χ^2-distance between the point and S, denoted by $d_i(S)$. The closeness of this profile to the subspace is then $m_i[d_i(S)]^2$; i.e., the squared distance weighted by the mass. The closeness of all the profiles to S is the sum of these quantities:

Formal definition of criterion optimized in CA

$$\text{closeness to } S = \sum_i m_i[d_i(S)]^2 \qquad (6.1)$$

The objective of CA is to discover the subspace S which minimizes this criterion. It can be shown that the subspace S being sought necessarily passes through the centroid of the points, as depicted in Exhibit 6.6, so we need to consider only subspaces that contain the centroid.

It is not necessary here to enter into the mathematical operations involved in this minimization. It suffices to say here that the most elegant way to define the theory of CA as well as to compute the solution to the above minimization problem is to use what is known in mathematics as the *singular value decomposition*, or SVD for short. The SVD is one of the most useful results in matrix theory, and has special relevance to all the methods of dimension reduction in statistics. It is to rectangular matrices what the eigenvalue–eigenvector decomposition is to square matrices, namely a way to break down a matrix into components, from the most to least important. The algebraic notion of *rank* of a matrix is equivalent to our geometric notion of dimension, and the SVD provides a straightforward mechanism of approximating a rectangular matrix with another matrix of lower rank by least squares. These results transfer directly into the theory of CA, and all the entities we need, the coordinates, principal inertias, etc., are obtained directly from the SVD. Since the SVD is available in many computing languages, the analytical part of CA is easily obtained. In the Computational Appendix we shall show how compactly CA can be programmed using the SVD function in the computing language R — see pages 219–220.

Singular value decomposition (SVD)

We have been describing the search for low-dimensional subspaces, for example, lines and planes, by least squares, and this sounds just like the objective of regression analysis, which also fits lines and planes to data points which can be imagined in multidimensional space. But there is a major difference between regression and what we are doing here. In regression one of the variables is regarded as a response variable and the distances that are minimized are parallel to the response variable axis. In our situation here, by contrast, there is no response variable, and fitting is done by minimizing the distances perpendicular to the subspace being fitted (see Exhibit 6.6 where the projections are perpendicular onto the plane, that is the shortest distances between the points and the plane). Sometimes fitting low-dimensional subspaces to points

Finding the optimal subspace is not regression

is referred to as "orthogonal regression", where the discovered dimensions are regarded as explanatory variables explaining the data points.

SUMMARY:
Reduction of
Dimensionality

1. Profiles consisting of m elements are situated exactly in spaces of dimensionality $m - 1$. Hence, profiles with more than four elements are situated in spaces of dimensionality greater than three, which we cannot observe directly.

2. If we can identify a subspace of lower dimensionality, preferably not more than two or three dimensions, which lies close to all the profile points, then we can project the profiles onto such a subspace and look at the profiles' projected positions in this subspace as an approximation to their true higher-dimensional positions.

3. What is lost in this process of dimensionality reduction is the knowledge of how far and in what direction the profiles lie "off" this subspace. What is gained is a view of the profiles that would not be possible otherwise.

4. The accuracy of display is measured by a quantity called the *percentage of inertia*. For example, if 85% of the inertia of the profiles is represented in the subspace, then the residual inertia, or error, which lies external to the subspace, is 15%.

5. The vertices, or unit profiles, can also be projected onto the optimal subspace. The object is not to represent the vertices accurately but to use them as reference points for interpreting the displayed profiles.

6. The actual computation of the low-dimensional subspace relies on measuring the closeness between a set of points and a subspace as the weighted-sum-of-squared χ^2-distances between the points and the subspace, where the points are weighted by their respective masses.

Optimal Scaling

So far CA has been presented as a geometric method of data analysis, stressing the three basic concepts of profile, mass and χ^2-distance, and the four derived concepts of centroid (weighted average), inertia, subspace and projection. Profiles are multidimensional points, weighted by masses, and distances between profiles are measured using the χ^2-distance. The profiles are visualized by projecting them onto a subspace of low dimensionality which best fits the profiles, and then projecting the vertex profiles onto the subspace as reference points for the interpretation. There are, however, numerous other ways to define and interpret CA and this is why the same underlying methodology has been rediscovered many times in different contexts. One of these alternative interpretations is called *optimal scaling* and a discussion of this approach at this point will provide additional insight into the properties of CA.

Contents

We refer again to the example in Exhibit 6.1, the cross-tabulation of age groups by self-perceived health categories. Both the row and column variables are categorical variables and are stored in a computer data file using codes 1 to 7 for age, and 1 to 5 for health. If we wanted to calculate statistics on the health variable such as mean and variance, or to use self-perceived health as a variable in a statistical analysis such as regression, it would be necessary to have values for each health category. It may not be true that each of the health categories is exactly one unit apart on such a scale, as is implicitly assumed if we use the values 1 to 5. The health categories are ordered (i.e., self-perceived health is an ordinal categorical variable), which

Quantifying a set of categories

49

indeed gives some minimal justification for using the values 1 to 5, but what if the variable were nominal, such as the country variable in Chapter 1 (see Exhibit 1.3) or a variable such as marital status?* The age group variable is also ordinal, established by defining intervals on the original age scale, so we could use the midpoints of each age interval as reasonable scale values, but it is not obvious what value to assign to age group 7, which is open-ended (75+ years). Failing any alternative, when categories are ordered as in this case, the integer values (1 to 7 and 1 to 5 here) are often used as default values in calculations. Optimal scaling provides a way of obtaining quantitative scale values for a categorical variable, subject to a specified criterion of optimality.

Computation of overall mean using integer scale

Initially we will use the default integer values in some simple calculations, but let us first reverse the coding of the health categories so that the higher value corresponds to better health — hence 5 indicates *very good* health, down to 1 indicating *very bad* health. In the data set as a whole, there are 817 respondents with *very good* health (code 5), 3542 with *good* health (code 4), and so on, out of a total sample of 6371 respondents. Using these integer codes as scale values for the health categories, the *average health category* in this sample can be calculated as follows:

$$[(817 \times 5) + (3542 \times 4) + ... + (103 \times 1)]/6371 = 3.72$$

i.e.

$$(0.128 \times 5) + (0.556 \times 4) + ... + (0.016 \times 1) = 3.72 \qquad (7.1)$$

where $817/6371 = 0.128$, $3542/6371 = 0.556$, etc. are the elements of the average row profile (see the last row of Exhibit 6.2). Therefore, this average across all the respondents is simply the weighted average of the scale values where the weights are the elements of the average profile.

Computation of group means using integer scale

Considering a particular age group now, say 16–24 years, we see from the first row of data in Exhibit 6.1 that there are 243 respondents with *very good* health, 789 with *good*, and so on, out of a total of 1223 in this young age group. Again, using the integer scale values 5 to 1 for the health categories, the *average health* for the 16–24 group is:

$$[(243 \times 5) + (789 \times 4) + ... + (6 \times 1)]/1223 = 4.02$$

i.e.

$$(0.199 \times 5) + (0.645 \times 4) + ... + (0.005 \times 7) = 4.02 \qquad (7.2)$$

where the second line again shows the profile values (for age group 16–24) being used as weights: $243/1223 = 0.199$, $789/1223 = 0.645$, etc. Thus we could say that the youngest age group has an average self-perceived health higher than the average: 4.02 compared to the average of 3.72. We could

---

* In my experience as a statistical consultant I once did see a survey with a variable "Religious Affiliation: 0=none, 1=Catholic, 2=Protestant, etc." and the researcher seriously calculated an "average religion" for the sample!

repeat the above calculation for the other six age groups and obtain averages as follows:

16–24	25–34	35–44	45–54	55–64	65–74	75+	*Overall*
4.02	3.97	3.86	3.66	3.39	3.30	3.19	3.72

Now that we have calculated health category means for the age groups using the integer scale values, we can compute the health category variance across the age groups. This is similar to the inertia calculation of Chapter 4 because each age group will be weighted proportional to its sample size. Alternatively you can think of all 6371 respondents being assigned the value corresponding to their respective age group, followed by the usual calculation of variance. The variance is calculated as (see Exhibit 6.1 for row totals):

Computation of variance using integer scale

$$\frac{1223}{6371}(4.02 - 3.72)^2 + \frac{1234}{6371}(3.97 - 3.72)^2 + \cdots + \frac{396}{6371}(3.19 - 3.72)^2 = 0.0857$$

with standard deviation $\sqrt{0.0857} = 0.293$.

The above calculations all depend on the initial use of the integer scale for the health categories, an arbitrary choice which is, admittedly, difficult to justify, especially after seeing the results of Chapter 6. The question is whether there are more justifiable, or at least more interesting, scale values. Answering this question depends on what is meant by "more interesting" and we now consider one possible criterion which leads to scale values related directly to CA. Let us suppose that the scale values for the health categories are denoted by the unknown quantities v_1, v_2, v_3, v_4 and v_5, to be determined. Then the average for all respondents would be, in terms of these unknowns, as in (7.1):

Calculating scores with unknown scale values

average health overall $= (0.128 \times v_1) + (0.556 \times v_2) + \cdots + (0.016 \times v_5)$ (7.3)

while the average for age group 16–24, for example, would be, as in (7.2):

average health 16–24 years $= (0.199 \times v_1) + (0.645 \times v_2) + \cdots + (0.005 \times v_5)$

(7.4)

The averages computed in this way are known as *scores*, so that (7.3) is the average score and (7.4) is the score, denoted by s_1, for the first age group. For each of the age groups, the score can be formulated in terms of the unknown scale values in the same way, leading to seven scores s_1, s_2, ..., s_7. Since each of the 6371 respondents is allocated to one of the age groups, each one can be associated with a corresponding score on the health scale. For example, all 1223 respondents in the 16–24 years age group would receive the score given by (7.4), so we can again imagine all 6371 respondents piling up at the seven different scores on the health scale, whatever these scores may be.

Now in order to determine the scale values, we can propose a property which we would like these 6371 scores to have. A desirable property is that the scores should be well separated so that the age groups be as distinct from one another as possible. Putting this the opposite way, it would be highly undesirable if

Maximizing variance gives optimal scale

the scores were so close to one another that it was difficult to distinguish between the different age groups in terms of their health categories. One way of phrasing this requirement more precisely is that we require the variance of the scores across all 6371 respondents to be a maximum. In numerical terms, we have 1223 respondents associated in the first age group (see first row of Exhibit 6.1) receiving score s_1, 1234 in the second age group receiving score s_2, and so on, and the variance is computed over all 6371 scores, just as we did on the previous page. Scale values v_1, v_2, $\ldots$, v_5 which lead to scores s_1, s_2, $\ldots$, s_7 with maximum variance will define what we call an *optimal scale*.

Optimal scale values from the best-fitting dimension of CA

Fortunately, it turns out that the positions of the health categories along the best-fitting CA dimension solve this optimal scaling problem exactly: the maximum variance is equal to the inertia on this optimal CA dimension, the coordinate values of the vertices in Exhibit 6.5 provide the optimal scale values, v_1 to v_5, and the coordinate values of the profiles provide the corresponding scores, s_1 to s_7. The actual coordinate values are given in Exhibit 7.1. We already know from Chapter 3 that an age group lies at the centroid of the five health category vertices, and this property carries over to any projection of the points onto a subspace. For example, if the profile of age group 16–24 (Exhibit 6.2) is used to weight the positions of the vertices of the five health categories (Exhibit 7.1), the following score is obtained:

$$(0.199 \times 1.144) + (0.645 \times 0.537) + \ldots + (0.005 \times -2.076) = 0.371$$

which agrees with the coordinate of the profile 16–24 in Exhibit 7.1.

Exhibit 7.1: Coordinate values of the points in Exhibit 6.5, i.e. the coordinates of the column vertices and the row profiles on the dimension that best fits the row profiles.

HEALTH CATEGORY	Vertex coordinate	AGE GROUP	Profile coordinate
Very good	1.144	16–24	0.371
Good	0.537	25–34	0.330
Regular	−1.188	35–44	0.199
Bad	−2.043	45–54	−0.071
Very bad	−2.076	55–64	−0.396
		65–74	−0.541
		75+	−0.658

The optimal scaling problem can be turned around by making a similar search for scale values for the age groups which maximize the variance of the health categories. The solution is given by the vertex coordinates for the age groups, and the scores for the health categories are their profile coordinates. The symmetry in the row and column problems is discussed further in the next chapter. This symmetry, or *duality*, of the scaling problems has led to calling the method *dual scaling*.

The optimal scale does not position the five health c[...]
tances from one another, like the original integer sca[...]
that there is a big difference between *good* and *regula*[...]
ference between *bad* and *very bad*. These scale values[...]
scores for the age groups that are the most separate[...]
ance criterion, in other words we have the maximum [...]
the age groups using the optimal scale for the health[...]
6.3, which displays only the age group scores, we can s[...]
changes in self-perceived health up to the age group 3[...]
large changes in the middle age categories, especially[...]
years, and then slower changes in the older groups.[...]
profile data in Exhibit 6.2, we can verify that from t[...]
group there is an approximate 50% drop in the *very goo*[...]
than doubling of the *bad* category, which accounts for t[...]
scores.

Handwritten note:

> optimal scale: maximize CANONICAL
> CORRELATION
> optimal score: BETWEEN ROW
> maximize variance & column.
> between M rows &/or.
> columns
>
> ---
>
> VERTEX coordinate = standard coord.
> Profile coordinate = ppal coordinate.
> BOOK: maximizing variance
> give optimal scale.
>
> ---
>
> IDENTIFICATION CONDITIONS or CONSTRAINTS
> Mean zero
> var = 1
> linear transformation still good.

The optimal health category scale values obtained are 1.144, 0.537, −1.188, −2.043 and −2.076 respectively (Exhibit 7.1). These numbers are calculated under certain restrictions which are required in order that a unique solution can be found. These restrictions are that the average on the health scale is 0 and the variance is 1 when applied to all 6371 respondents:

$$(0.128 \times 1.144) + (0.556 \times 0.537) + \ldots + (0.016 \times -2.076) = 0 \qquad \text{(mean 0)}$$

and

$$(0.128 \times 1.144^2) + (0.556 \times 0.537^2) + \ldots + (0.016 \times -2.076^2) = 1 \qquad \text{(variance 1)}$$

These prerequisites for the scale values are known as *identification conditions* or *constraints* in the jargon of mathematical optimization theory. The first condition is necessary since it is possible for two different sets of scale values to have different means but the same variance, so that it would be impossible to fix (or identify) a solution without specifying the mean. The second condition is required because if we arbitrarily multiplied the scale values by any large number, the variance of the eventual scores would be greatly increased as well — this would make no sense at all since we are trying to maximize the variance. Hence, it is necessary to look for a solution amongst scale values which have a fixed mean and fixed range of variation. The "mean 0, variance 1" condition is a conventional choice in such a situation, and conveniently leads to the vertex coordinates in CA, which satisfy the same conditions.

To determine the optimal scale, the two identification conditions described above are simply technical devices to ensure a unique mathematical solution to our problem. Having obtained the scale values, however, we are at liberty to transform them to a more convenient scale, as long as we remember that the mean and variance of the transformed scale are chosen for convenience and have no substantive or statistical relevance. The redefinition of this scale

Identification conditions for an optimal scale

Any linear transformation of the scale is still optimal

is usually performed by fixing the endpoints at some substantively meaningful values: for example, in the present case we could fix the *very bad* health category at 0 and the *very good* one at 100. So we need to make a transformation that takes the value of −2.076 to 0 and the 1.144 to 100. We can first add 2.076 to all five scale values, so that the lowest value is zero. The scale now ranges from 0 to 1.144 + 2.076 = 3.220. In order to have the highest value equal to 100, we then multiply all values by 100/3.220. So the formula in this particular case for computing a new scale value from the old one is simply

$$new = (old + 2.076) \times \frac{100}{3.220}$$

or, in the general case:

$$new = \left[(old - \text{old lower limit}) \times \frac{\text{new range}}{\text{old range}} \right] + \text{new lower limit} \qquad (7.5)$$

(in our example the new lower limit is zero). Applying this formula to all five optimal scale values results in the following transformed values:

Exhibit 7.2:
*Optimal scale values
from CA and the
values transformed
to lie between 0 and
100.*

HEALTH CATEGORY	*Optimal scale value*	*Transformed scale value*
Very good	1.144	100.0
Good	0.537	81.1
Regular	−1.188	27.6
Bad	−2.043	1.0
Very bad	−2.076	0.0

The previous 5-to-1 scale with four equal intervals between the scale points would have values 100, 75, 50, 25, 0 on the scale with range 100 (remember that we have reversed the scale so that *very good* is 100). The optimal transformed values show that *regular* is not at the midpoint (50) of the scale, but much closer to the "bad" end of the scale.

*Optimal scale
is not unique*

We should stress that the optimal scale depends on the criterion laid down for its determination as well as the chosen identification conditions. Apart from these purely technical issues, it clearly also depends on the particular cross-tabulation on which it is based. If we had a table which cross-tabulates health with another demographic variable, say education group, we would obtain a difference set of optimal scale values for the health categories, since they would now be optimally discriminating between the education groups.

*A criterion
based on
row-to-column
distances*

Finally, in contrast to the maximization criterion described above for optimal scaling, we present a minimization criterion for finding scale values which also leads to the CA solution, based on the distances from each row to each column — in the present example these will be distances between the health categories and age groups. First, imagine the health categories on any scale, for example the 1-to-5 integer scale from *very bad* to *very good* health, shown in Exhibit 7.3 on the next page. Then the objective is to find the positions of the age groups on the same scale so that they come as "close" as possible to the health categories in the sense that an age group that has higher frequency of a particular health category tends to be closer on the scale to that category. Suppose the health category values are $h_1, h_2,...,h_5$ (in this initial example, the values 1 to 5) and the age group values $a_1, a_2,...,a_7$. The distance between an age group and a health category is the absolute difference $|a_i - h_j|$, but we will prefer to use the squared distance $(a_i - h_j)^2$ as a measure of closeness[†]. To make distances count more depending on the frequency of occurrence in the cross-tabulation we will weight each squared distance by p_{ij}, the relative frequency as defined in Chapter 4, page 31, i.e., Exhibit 6.1 divided by its grand total 6371 (hence all the p_{ij}'s sum to 1). Our objective would then be to minimize the following function:

$$\sum_i \sum_j p_{ij} d_{ij}^2 = \sum_i p_{ij}(a_i - h_j)^2 \qquad (7.6)$$

showing that it pays more to make distances shorter when p_{ij} is higher. It is straightforward to show that, for any fixed set of health category points h_j, the minimum of (7.6) is achieved by the weighted averages for each age group. For the 1-to-5 scale values these weighted averages are just the set of scores we calculated before (see formula (7.2) and the set of scores shortly afterwards) which have also been depicted in Exhibit 7.3. But the positions

Exhibit 7.3:
The 1-to-5 scale of the health categories, and the weighted averages of the age groups.

of the two sets of points in Exhibit 7.3 minimize (7.6) given the fixed set of health categories, so the question is what the minimum would be over all possible configurations of scale values for the health categories. Again we need identification conditions for this question to make sense; otherwise the solution would simply put all health categories at the same point. If we add the same identification conditions that we had before, namely mean 0 and variance 1 for the scale values of the health categories, then the minimum is achieved by the optimal CA dimension once again. Comparing Exhibit 7.3 with the optimal positions in Exhibit 6.5, it is clear that the spread is higher in Exhibit 6.5, which in the unfolding interpretation means that all age group

[†] Again, as before, it is always easier to work with squared distances than distances — the square root in the Euclidean distance function causes many difficulties in optimization, and these disappear when we consider least-squares optimization.

points are closest to the health category points in terms of criterion (7.6). The value of the minimum achieved in Exhibit 6.5 is equal to 1 minus the (maximized) variance on the optimal CA dimension, and is sometimes referred to as the *loss of homogeneity* — we will return to this concept in Chapter 20 when discussing homogeneity analysis. Notice that the criterion (7.6) is easily generalized to two dimensions or more, say K dimensions, simply by replacing a_i and h_j by vectors of K elements and the squared differences $(a_i - h_j)^2$ by squared Euclidean distances in K-dimensional space.

SUMMARY:
Optimal Scaling

1. *Optimal scaling* is concerned with assigning scale values to the categories (or attributes) of a categorical variable to optimize some criterion which separates, or discriminates between, groups of cases, where these groups have been cross-tabulated with that variable.

2. The positions of the categories as vertex points on the optimal dimension of a CA provide optimal scale values in terms of a criterion that maximizes the variance between groups. The scores for the groups are the projections of their profiles on this dimension and the maximum variance is equal to the inertia of these projected profiles.

3. The coordinate positions of the projected categories on the optimal dimension are standardized in a certain way that is particular to the geometry of CA. For purposes of optimal scaling the mean and variance of the scale can be redefined; hence the scale values may be recentred and rescaled to conform to any scale convenient to the user, for example 0-to-1, or 0-to-100.

4. The optimal scale also satisfies a criterion based on the distances from each row point to each column point: that is, where the objective is to place the row and column points in a map such that the row-to-column distances, weighted by the frequencies in the contingency table, are minimized. This minimum is equal to 1 minus the maximum variance achieved in optimal scaling.

Symmetry of Row and Column Analyses

In all the examples and analyses presented so far, we have dealt with the analysis of the rows of a table, visualizing the row profiles and using the columns as reference points for the interpretation: let's call this the "row analysis". All this can be applied in a completely symmetric way to the columns of the same table. This can be thought of as transposing the table, making the columns the rows and vice versa, and then repeating all the procedures described in Chapters 2 to 7. In this chapter we shall show that the row analysis and column analysis are intimately connected. In fact, if a row analysis is performed, then the column analysis is actually being performed as well, and vice versa. CA can thus be regarded as the simultaneous analysis of the rows and columns.

Contents

Let us again consider the data in Exhibit 6.1 on self-perception of health. In Chapter 6 we performed the row analysis of these data since the object was to display the profiles of the age groups across the health categories. These seven profiles are contained in a four-dimensional space bounded by the five vertices which represent the extreme unit profiles corresponding to each health category. Then we diagnosed that most of the spatial variation of the profiles was along a straight line (Exhibit 6.3). The profiles were projected onto that line and the relative positions of these projections were interpreted as well as the projections of the five vertices (Exhibit 6.5).

Summary of row analysis

Exhibit 8.1:
Column profiles of health categories across the age groups, expressed as percentages.

AGE GROUP	Very good	Good	Regular	Bad	Very bad	Average
16–24	29.7	22.3	11.2	4.3	5.8	*19.2*
25–34	26.9	22.8	11.0	8.5	5.8	*19.4*
35–44	18.0	18.6	12.1	9.9	7.8	*16.2*
45–54	11.0	13.2	15.8	12.1	15.5	*13.5*
55–64	6.5	11.7	20.5	25.6	29.1	*14.3*
65–74	5.4	7.5	19.0	23.7	19.4	*11.2*
75+	2.4	3.8	10.5	15.9	16.5	*6.2*
Sum	*100*	*100*	*100*	*100*	*100*	*100*

Column analysis — profile values have symmetric interpretation

We now consider the alternative problem of displaying the column profiles of Exhibit 6.1, i.e. the profiles of the health categories across the size classes, shown in Exhibit 8.1. The column profiles give, for each health category, the percentages of respondents across the age groups: for example, in the *bad* health category, 4.3% are 16–24 years, 8.5% 25–34 years, and so on. Although this table of column profiles looks completely different to the row profiles in Exhibit 6.2, when we look at specific values and compare them to their averages we can see that they contain the same information (we already noticed this in Chapter 2, for the travel data set). For example, consider the value in the *bad* column for the age group 65–74 years: 23.7%. Compare this value with the average percentage of *bad* in the whole sample, given in the last column: 11.2%. Thus we conclude that in the 65–74 age group there are just over twice as many respondents saying their health is *bad* compared to the overall average — in fact, the ratio is $23.7/11.2 = 2.1$. If we look at the same cell of Exhibit 6.2, we see that, of the 65–74 year olds, 13.7% assess their health as *bad*, while the proportion of 65–74 year olds in the sample is 6.5% (last row of Exhibit 6.2). Hence again, just over twice as many compared to the average, say their health is *bad* and in fact the ratio is identical: $13.7/6.5 = 2.1$.

Column analysis — same total inertia

In Chapter 4 it was shown that the total inertia of the column profiles is equal to the total inertia of the row profiles — the two calculations are just alternative ways of writing the same formula, the χ^2 statistic divided by the sample size. For the health perception data, the total inertia is equal to 0.1404.

Column analysis — same dimensionality

The column profiles define a cloud of five points, each with seven components, which should then lie in a space of dimension six, using the same argument as before because the components add up to 1. It turns out, however, that five points do not even fill all six dimensions of this space, but only four of the dimensions. One way to grasp this fact intuitively is to realize that two points lie exactly on a one-dimensional line, three points lie in a two-dimensional plane, four points lie in a three-dimensional space, and so five points lie in a four-dimensional space. Hence, although the row profiles and column profiles lie in different spaces, the dimensionality of these two clouds of points is identical, in this case it is equal to four. This is the first geometric way in which the analyses of the row and of the column profiles are the same. Many more similarities will soon become apparent.

Still considering the five health category profiles in four-dimensional space, we now ask the same question as before: Can these points be approximately displayed in a lower-dimensional subspace and what is the quality of this approximation? By performing an analogous set of mathematical calculations as was required in Chapter 6, it turns out that the column profiles are well approximated by a one-dimensional line, and the quality of the approximation is 97.3%, exactly the same percentage that was obtained in the case of the row profiles. This is the second geometric property which is identical in the two analyses.

Column analysis — same low-dimensional approximation

Exhibit 8.2: *Optimal one-dimensional map of the health category profiles.*

The projections of the column profiles onto the best-fitting line are shown in Exhibit 8.2. Here we see that the health categories lie in an order which concurs exactly with the positions of the vertices in Exhibit 6.5. The actual values of their coordinate positions are not the same but the relative positions are identical. According to the scale of Exhibit 8.2 and comparing the positions of the health categories in Exhibit 6.5, it appears that the coordinates of the profiles are a contracted, or shrunken, version of the vertices. A specific interpretation of this "contraction factor" will be given soon. Furthermore, in Exhibit 8.3 the projections onto this line of the seven outer vertices, representing the age groups, are displayed. Comparing the positions of the

Column analysis — same coordinate values, rescaled

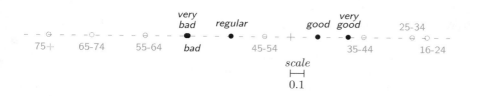

Exhibit 8.3: *Same map as Exhibit 8.2, showing the projected positions of the age group vertices.*

vertices here with those of the age group profiles in Exhibit 6.5 (or Exhibit 6.3 where the scale is larger) reveals exactly the same result for the rows — the positions of the row profiles with respect to their best-fitting line in Exhibit 6.5 are a contracted version of the positions of the age group vertices projected onto the best-fitting line of the health category profiles in Exhibit 8.2. Putting this the opposite way, the positions of the row vertices in the column analysis are a simple expansion of the positions of the row profiles in the row analysis. This is the third and most important way in which the two analyses are related geometrically.

Principal axes and principal inertias

The best-fitting line in each analysis is called a *principal axis*. More specifically it is referred to as the "first principal axis", since there are other principal axes, as we shall see in the following chapters. We have seen that in both row and column analyses the total inertia is equal to 0.1404 and that the percentage of inertia accounted for by the first axis is 97.3%. The specific part of inertia that is accounted for by the first axis is equal to 0.1366 in both cases, which gives the percentage explained as $100 \times 0.1366/0.1404 = 97.3\%$. The inertia amount (0.1366) accounted for by a principal axis is called a *principal inertia*, in this case the first principal inertia because it refers to the first principal axis. It is also often called an *eigenvalue* because of the way it can be calculated, as an eigenvalue of a square symmetric matrix.

Scaling factor is the square root of the principal inertia

It seems, then, that we have to do only one analysis — either the row analysis or the column analysis. The results of the one can be obtained from those of the other. But what is the exact connection between the two; in other words what is the scaling factor which can be used to pass from vertex positions in one analysis to profile positions in the other? This scaling factor turns out to be equal to the square root of the principal inertia itself; e.g., in this example it is $\sqrt{0.1366} = 0.3696$. Thus to pass from the row vertices in Exhibit 8.3 to the row profiles in Exhibits 6.3 or 6.5, we simply multiply the coordinate values by 0.3696, i.e., just over one-third. Conversely, to pass from the column profiles in Exhibit 8.3 to the column vertices in Exhibit 6.5, we multiply the coordinate values by the inverse, namely $1/0.3696 = 2.706$. The numerical values of all the profile and vertex coordinates are given in Exhibits 7.1 and 8.4, and the following simple relationship for both rows and columns can easily be verified comparing these two exhibits:

$$\text{profile coordinate} = \text{vertex coordinate} \times \sqrt{\text{principal inertia}}$$

Notice in Exhibits 6.5 and 8.2 that the profile points are more bunched up than the vertex points. The scaling factor is a direct measure of how bunched up the "inner" profiles are compared to the "outer" vertices. In this case, the scaling factor of 0.3696 implies that the spread of the profiles is about one third that of the vertices. At the end of Chapter 4 the total inertia was interpreted as the amount of dispersion in a set of profiles relative to the outer vertices (see Exhibit 4.1). The principal inertias (or their square roots

HEALTH CATEGORY	*Profile coordinate*
Very good	0.423
Good	0.198
Regular	−0.439
Bad	−0.755
Very bad	−0.767

AGE GROUP	*Vertex coordinate*
16–24	1.004
25–34	0.893
35–44	0.538
45–54	−0.192
55–64	−1.070
65–74	−1.463
75+	−1.782

Exhibit 8.4:
Coordinate values of the points in Exhibit 8.2, i.e. the coordinates of the column profiles and the row vertices on the first principal axis of the column profiles (cf. Exhibit 7.1).

which we are considering here) are also measures of dispersion but refer to individual principal axes rather than to the whole profile space. The higher the principal inertia is, and thus the higher the scaling factor is, the more spread out the profiles are relative to the vertices, along the respective principal axis. It should now be obvious that a principal inertia can not be greater than 1 — the profiles must be in positions "interior" to their corresponding vertices.

Correlation interpretation of the principal inertia

The square root of the principal inertia, which as we already pointed out is always less than 1, has an alternative interpretation as a correlation coefficient. A correlation coefficient is usually calculated between pairs of measurements, for example the correlation between income and age. In the present example there are two observations on each respondent — age group and health category — but these are categorical observations, not measurements. A correlation coefficient between these two variables can be computed using the default integer codes of 1 to 7 for the age groups and 1 to 5 for the health categories. The correlation is then computed to be 0.3456. Using any other set of scale values would give a different correlation, so the following question arises: Which scale values can we assign to the age groups and health categories such that the correlation is the highest? The maximum correlation found in this way is sometimes called the *canonical correlation*. In the present example, the canonical correlation turns out to be 0.3696, exactly the square root of the principal inertia, i.e., the scaling factor linking the row and column analyses. The scale values for the age groups and the health categories that yield this maximum correlation are just the coordinate values of the age groups and size classes on the first CA principal axis, given in Exhibits 7.1 and 8.3 and displayed in Exhibits, 6.3, 6.5, 8.2 and 8.3. We can use profile or vertex coordinates, since correlation is unaffected by recentring or rescaling the scale values. It is conventional to use standardized scale values, with mean 0 and variance 1, to identify the solution uniquely.

Graph of the correlation

A correlation between two variables is usually illustrated graphically by a scatterplot of the cases, e.g., age group (*y*-axis) by health category (*x*-axis). Although we have 6371 cases in the scatterplot, there are only 7 values along

Exhibit 8.5:
Scatterplot according to the scale values which maximize the correlation between health category and age group; squares are shown at each combination of values, with area proportional to the number of respondents. The correlation is equal to 0.3456.

the y axis and 5 values along the x axis, thus only $7 \times 5 = 35$ possible points in this scatterplot (Exhibit 8.5). At a specific point corresponding to a health category and age group lie all the cases in the respective cell of the original cross-table (Exhibit 6.1), displayed here in the form of a square with an area proportional to the cell frequency. The canonical correlation is then the usual Pearson correlation of all 6371 cases in this scatterplot. The optimal property of the canonical correlation means that there is no other way of scaling the row and column categories which would yield a higher correlation coefficient in such a scatterplot. A canonical correlation of 1 would be attained if all points were lying on a straight line, which means that each age group is associated with only one health category (i.e., the profiles are all unit profiles, or vertex points).

Principal coordinates and standard coordinates

It is convenient to introduce some terminology at this stage to avoid constant repetition of the phrases "coordinate positions of the vertices" and "coordinate positions of the profiles". Since the former coordinates are standardized to have mean 0 and variance 1, we call them *standard coordinates*. Since the latter coordinates refer to the profiles with respect to principal axes, we call them *principal coordinates*. For example, the first column of numerical results in Table 8.2 contains the principal coordinates of the health categories (columns), while the second column contains the standard coordinates of the age groups.

In both cases these are coordinates on the first principal axis of the CA; in future chapters we shall usually have more than one principal axis.

Another way of thinking about the correlation definition of CA is that each of the 6371 individuals in the health survey example can be assigned a pair of scale values, one (a_i, say) for age group and one (h_j) for health category. As before, these scale values are unknown but we define a criterion to optimize in order to determine them. Each individual has a score equal to the sum of these two scale values, $a_i + h_j$; for example, someone in the **25–34** age group with *very good* health (second age group and first health category) would have a score of $a_2 + h_1$. Suppose that the correlation of the pairs of values $\{a_i, a_i + h_j\}$ is denoted by $\mathrm{cor}(a, a + h)$, where a and h denote the 6371 scale values for the whole sample; similarly, the correlation for the pairs $\{h_j, a_i + h_j\}$ is denoted by $\mathrm{cor}(h, a + h)$. A criterion to optimize would be to find the scale values that optimize these two correlations in some way. It can be shown that the first dimension of the CA solution gives scale values which are optimal in the sense that they maximize the average of the squares of these correlations:

$$\text{average squared correlation} = \frac{1}{2}[\mathrm{cor}^2(a, a + h) + \mathrm{cor}^2(h, a + h)] \qquad (8.1)$$

Since $\mathrm{cor}(X, X + Y) = \frac{1}{2}[1 + \mathrm{cor}(X, Y)]$ for any two variables X and Y, the average squared correlation in (8.1) is equal to:

$$\text{average squared correlation} = \frac{1}{4}[1 + \mathrm{cor}^2(a, h)]^2 \qquad (8.2)$$

Therefore, when the CA solution maximizes $\mathrm{cor}(a, h)$, i.e., the canonical correlation, it also maximizes (8.2), equivalently (8.1). This result will be useful later because it can easily be generalized to more than two variables — see Chapter 20.

Yet another equivalent way of expressing the optimality of the CA solution is as follows, using the notation of the previous section. Instead of sums of scale values, calculate an average for each person: $\frac{1}{2}(a_i + h_j)$. Then calculate the differences between each person's age value and health value and the average: $a_i - \frac{1}{2}(a_i + h_j)$ and $h_j - \frac{1}{2}(a_i + h_j)$. A measure of how similar the age values are to the health values is the sum of squares of these two differences, which we average as well, leading to a measure of variance of the two values a_i and h_j):

Minimizing loss of homogeneity within variables

$$\text{variance (for one case)} = \frac{1}{2}\left([a_i - \frac{1}{2}(a_i + h_j)]^2 + [h_j - \frac{1}{2}(a_i + h_j)]^2\right) \qquad (8.3)$$

The term *homogeneity* is used in this context because if the scale values a_i and h_j are the same, their variance would be zero; hence an individual with this combination of categories is called *homogeneous*. An alternative term for homogeneity is *internal consistency*. Averaging the values (8.3) for the whole sample, we obtain an amount which is called the *loss of homogeneity* (see page 56, where this term is used in the same sense). If all the age values coincided with the health values, the loss of homogeneity would be zero, that is the sample would be completely homogeneous (or internally consistent). The aim is to find scale values which minimize this loss, and once more the solution coincides with the coordinates of the age and health points on the first CA dimension. Again it is clear that this definition is easily extended to more than two variables, as we will do in Chapter 20.

SUMMARY:
Symmetry of Row
and Column
Analyses

1. Everything we did in the row analysis can be applied in a completely symmetric fashion to the columns, as if the table were transposed and all operations repeated.

2. The column analysis thus leads to the visualization of the column profiles in their optimal subspace of display, along with the display of the vertices representing the rows.

3. The best-fitting line, or dimension, is called the (first) *principal axis* of the profiles. The amount of inertia this dimension accounts for is called the (first) *principal inertia*.

4. The coordinate positions of profiles with respect to a principal axis are called *principal coordinates* and the coordinate positions of vertices with respect to a principal axis are called *standard coordinates*.

5. The two analyses are equivalent in the sense that each has the same total inertia, the same dimensionality and the same decomposition of inertia into principal inertias along principal axes.

6. Furthermore, the profiles and vertices in the two analyses are intimately related as follows: along a principal axis profile positions (in principal coordinates) have exactly the same relative positions as the corresponding vertices (in standard coordinates) in the other analysis, but are reduced in scale. The scaling factor involved is exactly the square root of the principal inertia along that axis.

7. This scaling factor can also be interpreted as a *canonical correlation*, especially when we are dealing with the first principal axis. It is the maximum correlation that can be attained between the row and column variables as a result of assigning numerical quantifications to the categories of these variables.

Two-dimensional Displays

We have discussed at some length the projections of a cloud of profiles onto a single principal axis, the best-fitting straight line. In practice you will find that most of the reported CA displays are two-dimensional, usually with the first principal axis displayed horizontally (the x-axis) and the second principal axis vertically (the y-axis). In general, the projections may take place onto any low-dimensional subspace, but the two-dimensional case is, of course, rather special because of our two-dimensional style of displaying graphics on computer screens or on paper. In the Computational Appendix there are also some examples using the R programming language to do CA graphics in three dimensions (e.g., Exhibit B.5 on page 235).

Contents

The next example, which appeared originally in my 1984 book *Theory and Applications of Correspondence Analysis*, has been adopted as a test example in almost all the implementations of CA in the major commercial statistical packages. This example still serves as an excellent introduction to two-dimensional displays and has also been referred to in several journal articles, even though it is an artificial data set. It concerns a survey of all 193 staff members of a company, in order to formulate a smoking policy. The staff members are cross-tabulated according to their rank (five levels) and a categorization of their smoking habits (four groups) — the contingency table is reproduced in Exhibit 9.1. Because it is a 5×4 table, its row profiles and column profiles lie exactly in three-dimensional spaces.

Data set 4: Smoking habits of staff groups

Exhibit 9.1:
*Cross-tabulation of
staff group by
smoking category,
showing row profiles
and average row
profile in
parentheses, and the
row masses.*

| STAFF | SMOKING CATEGORIES | | | | Row | |
GROUPS	None	Light	Medium	Heavy	Totals	Masses
Senior managers	4	2	3	2	*11*	*0.057*
SM	(0.364)	(0.182)	(0.273)	(0.182)		
Junior managers	4	3	7	4	*18*	*0.093*
JM	(0.222)	(0.167)	(0.389)	(0.222)		
Senior employees	25	10	12	4	*51*	*0.279*
SE	(0.490)	(0.196)	(0.235)	(0.078)		
Junior employees	18	24	33	13	*88*	*0.456*
JE	(0.205)	(0.273)	(0.375)	(0.148)		
Secretaries	10	6	7	2	*25*	*0.130*
SC	(0.400)	(0.240)	(0.280)	(0.080)		
Total	*61*	*45*	*62*	*25*	*193*	
Average Profile	*(0.316)*	*(0.233)*	*(0.321)*	*(0.130)*		

Row analysis   As before, this table may be thought of as a set of rows or a set of columns. We assume that the row analysis is more relevant; that is, we are interested in displaying for each staff group what percentage are non-smokers, what percentage are light smokers, and so on. The row profile space is a four-pointed simplex, a tetrahedron, in three dimensions, which is three-dimensional equivalent of the triangular space previously (this can be seen using the three-dimensional graphics described in the Computational Appendix). To reduce the dimensionality of the profiles, they should be projected onto a best-fitting plane (see Exhibit 6.6 on page 46). The map, shown in Exhibit 9.2, also shows the projections of the four vertex points representing the smoking groups. Notice that the first principal axis customarily defines the horizontal axis of the map, and the second principal axis the vertical axis. On the axes the respective principal inertias are given (0.07476 and 0.01002 respectively), as well as the corresponding percentages of inertia. These values can be accumulated to give the amount and percentage of inertia accounted for by the plane of the two axes. Thus the inertia in the plane is 0.08478, which is 99.5% of the total inertia of 0.08519. This means that by sacrificing one dimension we have lost only 0.5% of the inertia of the profile points. Putting this another way, the five row profiles lie very close to this plane of representation, so close that we can effectively ignore their distance from the plane when exploring their relative positions.

*Interpretation
of row profiles and
column vertices*   Looking only at the profiles' positions for a moment, we can see that the groups furthest apart are Junior Employees (JE) and Junior Managers (JM) on the left-hand side, opposed to Senior Employees (SE) on the right-hand side — hence the greatest differences in smoking habits are between these extremes. Senior Managers (SM) appear to lie between Junior Managers and

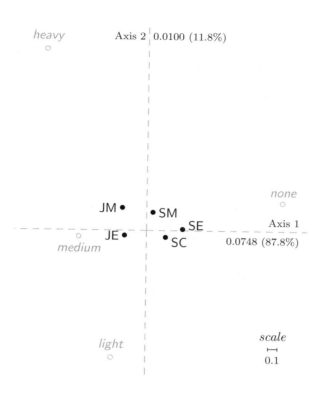

Exhibit 9.2:
Optimal two-dimensional CA map of the smoking data of Table 9.1, with rows in principal coordinates (projections of profiles) and columns in standard coordinates (projections of vertices).

Senior Employees, while Secretaries (SC) are quite close to Senior Employees. In order to explain the similarities and differences between the staff groups, it is necessary to inspect the positions of the profiles relative to the vertices. Since the three smoking categories are on the left and the non-smoking category is on the right, the left-to-right distinction is tantamount to smokers versus non-smokers. The groups JE and JM are different to SE because the former groups have relatively more smokers, and SE has relatively more non-smokers. The centre of such a display is always the average profile, so that we can also consider the deviations of the staff groups outwards from the average profile in different directions, the main deviations being from left to right.

The two-dimensional display is such that it actually contains the best one-dimensional display in it as well. If all the points in Exhibit 9.2 were projected vertically onto the horizontal axis, then this unidimensional display would be the one obtained by looking for the best one-dimensional display right from the start. The principal axes are said to be *nested*, in other words an optimal display of a certain dimensionality contains all the optimal displays of lower dimensionality. Notice that the three smoking groups on the left will project very close together on the first axis, a long way from the non-smoking point

Nesting of principal axes

on the right. This is the greatest single feature in the data. Putting this in the optimal scaling terminology of Chapter 7, a "smoking scale" which best differentiates the five staff groups is not one which assumes equal intervals between the four smoking categories, but rather one which places all three smoking categories quite close to one another but far from the non-smoking category, effectively a dichotomous smoking/non-smoking dichotomy.

Interpretation of second dimension

Continuing with the two-dimensional interpretation, we see that the second (vertical) principal axis pulls apart the three smoking levels. The profiles do not differ as much vertically as horizontally, as indicated by the much lower percentage of inertia on the second axis. Nevertheless, we can conclude that the profile of JE has relatively more light smokers than heavy smokers compared to that of JM, even though both these groups have similar percentages of smokers as seen by their similar positions on the horizontal axis. These conclusions can be easily verified in the original data of Exhibit 9.1.

Verifying the profile–vertex interpretation

One way of verifying the interpretation of the positions of the profiles relative to the vertices is to measure the profile-to-vertex distances in Exhibit 9.2 and then compare these to the profile values. This verification should be performed one vertex at a time, for example the five distances from the staff groups to the vertex *light*. As a general rule, assuming that the display is of good quality, which is true in this case, the closer a profile is to that vertex, the higher its profile is for that category. For example, an interpretation which we made in the previous paragraph is that because JE lies more towards *light* than JM, JE should have relatively more light smokers than JM. The actual data are that 24/88 or 27% of JEs are light smokers, whereas 3/18 or 17% of JMs are light smokers, so this agrees with our interpretation. Exhibit 9.3 graphically compares all profile-to-vertex distances to their corresponding profile values. The abbreviation 42, for example, is used for JE-to-*light* (row 4, column 2) and sf 22 for JM-to-*light* (row 2, column 2). Clearly the higher profile element of 0.27 for 42 corresponds to a smaller distance than the profile of 0.17 for 22. For each vertex, we say that the profile elements are *monotonically inversely* related to the profile-to-vertex distances, which in graphical terms means that each set of five points in Exhibit 9.3 corresponding to a particular vertex forms a descending pattern from top left to bottom right. For example, the set of five points corresponding to the fourth vertex point (*heavy*), with labels 34, 54, 44, 14 and 24, are arranged in such a descending sequence.

Asymmetric maps

We say that the Exhibit 9.2 is an *asymmetric map*, or a map which is *asymmetrically scaled*, because it is the joint display of profile and vertex points. In an asymmetric map, therefore, one of the sets of points, in this case the rows, is scaled in principal coordinates, while the other is scaled in standard coordinates. If we were more interested in the column analysis, then the column points would be in principal coordinates and the row points in standard coordinates. What we said in Chapter 8 about the scaling factor between the row

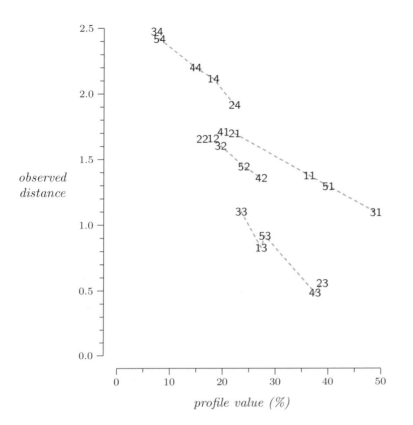

Exhibit 9.3:
*The measured profile-to-vertex distances in Exhibit 9.2 plotted against the corresponding values of the row profiles of Exhibit 9.1. Each row-column pair is labelled with their respective category numbers: for example, row profile 3 (senior employees) and column vertex 4 (heavy smoking) are denoted by **34**. Notice the descending pattern with increasing profile value for each set of distances corresponding to a particular vertex, with some small exceptions.*

and column problems holds for each principal axis. Thus the two-dimensional display of the column profiles would be a shrunken version of the positions of the column vertices given in Exhibit 9.2, but the "shrinking factors" (i.e., the canonical correlations, equal to the square roots of the principal inertias) along the two axes are not the same: $\sqrt{0.07476} = 0.273$ and $\sqrt{0.01002} = 0.100$ respectively. Thus along the first axis the shrinking is by a factor of 0.273 (i.e., just under four-fold) and along the second axis by a factor of 0.1 (i.e., ten-fold). By the same argument, to pass from the row profiles in Exhibit 9.2 to their vertex positions in the column problem we would simply expand them nearly four-fold along the first axis and ten-fold along the second axis. Apart from these scaling factors the relative positions of the profiles and the vertices are the same. Exhibit 9.4 shows the other possible asymmetric map, where the columns are represented as profiles in principal coordinates and the rows as vertices in standard coordinates. In this map the column points are at weighted averages of the row points using the elements of the column profiles as weights. The asymmetric map of Exhibit 9.2 is often called the *row principal* map (because row points are in principal coordinates) and Exhibit 9.4 the *column principal* map.

Exhibit 9.4:
*Asymmetric CA
map of the smoking
data of Table 9.1,
with columns in
principal
coordinates and
rows in standard
coordinates.*

Symmetric map   Having gone to great lengths to explain the geometry of asymmetric displays, we now introduce an alternative way of mapping the results, called the *symmetric map*. This option is by far the most popular in the CA literature, especially amongst French researchers. In a symmetric map the separate configurations of row profiles and column profiles are overlaid in a joint display, even though they emanate, strictly speaking, from different spaces. In a symmetric map, therefore, both row and column points are displayed in principal coordinates. Exhibit 9.5, for example, is the symmetric map of the smoking data, and is thus an overlay of the two sets of "inner" points in black in Exhibits 9.2 and 9.4. This simultaneous display of rows and columns finds some justification in the intimate relationship between the row and column analyses, involving a simple scaling factor between profiles and corresponding vertices. The convenience of such a display is that, whatever the absolute level of association might be, we always have both clouds of points equally spread out across the plotting area, hence there is less possibility of overlapping labels in the display. In asymmetric maps, by contrast, the profile points (which are usually the points of primary interest) are often bunched up in the middle of the display, far from the outer vertices and the visualization is generally less aesthetic.

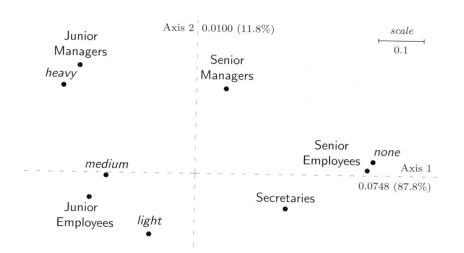

Exhibit 9.5:
*Symmetric map of
smoking data; both
rows and columns
are in principal
coordinates.*

Since both clouds of profiles are displayed simultaneously in Exhibit 9.5, the plotted row-to-row distances approximate the inter-row χ^2 distances and the plotted column-to-column distances approximate the inter-column χ^2-distances. Of course, the inter-row distance interpretation applies to the points in Exhibit 9.2 as well, since this is the identical display of the rows which is used in Exhibit 9.5 (note the difference in scales between these two maps) — similarly for the column points in Exhibit 9.4. The interpoint χ^2-distances can be verified by plotting the observed distances versus the true ones, as in Exhibit 9.6. There is an excellent agreement, which was to be expected since the quality of display of the profiles is 99.5% in both cases.

*Verification of
interpoint
chi-squared
distances in
symmetric map*

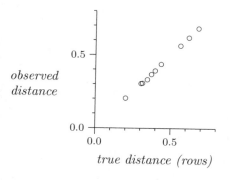

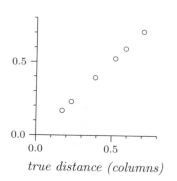

Exhibit 9.6:
*Observed interpoint
row distances and
interpoint column
distances measured
in Exhibit 9.5,
plotted against the
true χ^2-distances
between the row
profiles and between
the column profiles,
respectively, of
Table 9.1.*

Danger in interpreting row-to-column distances in a symmetric map

There is a price to pay for the convenience of the symmetric map which comes in the form of a danger in interpreting row-to-column distances directly. No such distance is defined or intended in this map. This is the aspect of CA that is often misunderstood and has caused some confusion amongst users who would like to make clusters of row and column points in a symmetric map (see the Epilogue, page 267). Strictly speaking, it is not possible to deduce from the closeness of a row and column point the fact that the corresponding row and column necessarily have a high association. Such an interpretation is justified to a certain extent in the case of the asymmetric map, as illustrated in Exhibit 9.5. A golden rule in interpreting maps of this type is that interpoint distances can be interpreted whenever the points concerned are situated in the same space, for example row profiles along with the vertex points representing the columns in the row profile space. When interpreting a symmetric map, the fact that this is the overlay of two separate maps should always be borne in mind. In Chapter 13 the row–column interpretation called the "biplot" will be described — this is the more accurate way of thinking about the joint display of rows and columns.

SUMMARY: Two-dimensional Displays

1. As the dimensionality of the subspace of display is increased, so the capacity of the display to represent the profile points accurately is improved. There is, however, a trade-off in the sense that the visualization of the points becomes more and more complex beyond two dimensions. Two-dimensional displays are usually the displays of choice.

2. The principal axes are *nested*; i.e., the first principal axis found in the one-dimensional solution is identical to the first principal axis in the two-dimensional solution, and so on. Increasing the dimensionality of the display simply implies adding new principal axes to those already found.

3. An *asymmetric map* is one in which the row and column points are scaled differently, e.g., the row points in principal coordinates (representing the row profiles) and the column points in standard coordinates (representing the column vertices). There are thus two asymmetric plots possible, depending on whether the row or column analysis is of chief interest.

4. In an asymmetric map where the rows, for example, are in principal coordinates (i.e., the row analysis), distances between displayed row points are approximate χ^2-distances between row profiles; and distances from the row profile points to a column vertex point are, as a general rule, inversely related to the row profile elements for that column.

5. A more common type of display, however, is the *symmetric map* where both rows and columns are scaled in principal coordinates.

6. In a symmetric map, the row-to-row and column-to-column distances are approximate χ^2-distances between the respective profiles. There is no specific row-to-column distance interpretation in a symmetric map.

Three More Examples

To conclude these first 10 introductory chapters to the CA of a two-way table, we now give three additional examples: (i) a table which summarizes the classification of scientists from ten research areas into different categories of research funding; (ii) a table of counts of 92 marine species at a number of sampling points on the ocean floor; (iii) a linguistic example, where the letters of the alphabet have been counted in samples of texts by six English authors. In the course of these examples we shall discuss some further issues concerning two-dimensional displays, such as the interpretation of dimensions, the difference between asymmetric and symmetric maps, and the importance of the aspect ratio of the map.

Contents

The data come from a scientific research and development organization which has classified 796 scientific researchers into five categories for purposes of allocating research funds (Exhibit 10.1). The researchers are cross-classified according to their scientific discipline (the 10 rows of the table) and funding category (the five columns of the table). The categories are labeled *A*, *B*, *C*, *D* and *E*, and are in order from highest to lowest categories of funding. Actually, *A* to *D* are the categories for researchers who are receiving research grants, from *A* (most funded) to *D* (least funded), while *E* is a category assigned to researchers whose applications were not successful (i.e., funding application rejected).

Data set 5: Evaluation of scientific researchers

Exhibit 10.1:
*Frequencies of
funding categories
for 796 researchers
who applied to a
research agency: A is
the most funded, D
is the least funded,
and E is not funded.*

SCIENTIFIC AREAS	*FUNDING CATEGORIES*					
	A	*B*	*C*	*D*	*E*	*Sum*
Geology	3	19	39	14	10	*85*
Biochemistry	1	2	13	1	12	*29*
Chemistry	6	25	49	21	29	*130*
Zoology	3	15	41	35	26	*120*
Physics	10	22	47	9	26	*114*
Engineering	3	11	25	15	34	*88*
Microbiology	1	6	14	5	11	*37*
Botany	0	12	34	17	23	*86*
Statistics	2	5	11	4	7	*29*
Mathematics	2	11	37	8	20	*78*
Sum	*31*	*128*	*310*	*129*	*198*	*796*
Average Row Profile	*3.9%*	*16.1%*	*38.9%*	*16.2%*	*24.9%*	

*Decomposition
of inertia*

This 10×5 table lies exactly in four-dimensional space and the decomposition of inertia along the four principal axes are as follows:

Dimension	Principal inertia	Percentage of inertia
1	0.03912	47.2%
2	0.03038	36.7%

Each axis accounts for a part of the inertia, expressed as a percentage. Thus the first two dimensions account for almost 84% of the inertia. The sum of the principal inertias is 0.082879, so the χ^2 statistic is $0.082879 \times 796 = 65.97$. If one wants to perform the statistical test using the χ^2 distribution with $9 \times 4 = 36$ degrees of freedom, this value is highly significant ($P = 0.002$).

*Asymmetric
map of row profiles*

Exhibit 10.2 shows the asymmetric map of the row profiles and the column vertices. In this display we can see that the magnitude of the association between the disciplines and the research categories is fairly low; in other words the profiles do not deviate too much from the average (cf. Exhibit 4.2). This situation is fairly typical of social science data, so the asymmetric map is not so successful because all the profile points are bunched up in the middle of the display — in fact, they are so close to one another that we cannot write the full labels and have just put the first two letters of each discipline. Nevertheless, we can interpret the space easily looking at the positions of the vertices. The horizontal dimension lines up the four categories of funding in their inherent ordering, from *D* (least funded) to *A* (most funded), with *B* and *C* close together in the middle. The vertical dimension opposes category *E* (not funded) against the others, so the interpretation is fairly straightforward. The more a discipline is high up in this display the less its researchers are actually granted funding. The more a discipline lies to the right of this display, the more funding its funded researchers receive. Using marketing research terminology,

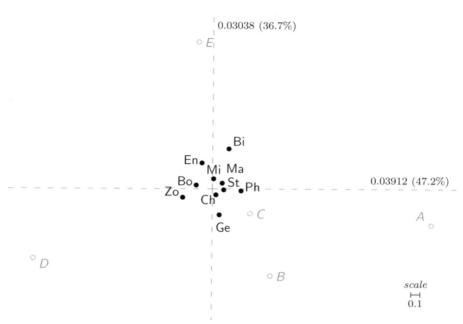

Exhibit 10.2:
*Asymmetric map of
the row profiles of
Table 10.1 (scientific
funding data).*

the "ideal point" is in the lower right of the map: more grant applications accepted (low down), and those accepted receiving good classifications (to the right). Hence, if we were doing a trend study over time, disciplines would need to move towards the bottom right-hand side to show an improvement in their funding status. At the moment there are no disciplines in this direction, although Physics is the most to the right (highest percentage — 10 out of 114, or 8.8% — of type *A* researchers), but is at the middle vertically since it has a percentage of non-funded researchers close to average (26 out of 114 not funded, or 22.8%, compared to the average of 198 out of 796, or 26.5%).

Exhibit 10.3 shows the symmetric map of the same data, so that the only difference between this display and that of Exhibit 10.3 is that the column profiles are now displayed rather than the column vertices, leading to a change in scale which magnifies the display of the row profiles. This zooming in on the configuration of disciplines facilitates the interpretation of their relative positions and also gives space for fuller labels. The relative positions of the disciplines can now be seen more easily: for example, Geology, Statistics, Mathematics and Biochemistry are all at a similar position on the first axis, but widely different on the second. This means that the researchers in thse fields whose grants have been accepted have similar positions with respect to the funded categories *A* to *D* categories, but Geology has much fewer rejections (11.8% of category *E*) than Biochemistry (41.4%). In this symmetric display we cannot assess graphically the overall level of association (inertia) between the rows and the columns. This can be assessed only from the numerical

Symmetric map

Exhibit 10.3:
Symmetric map of
Table 10.1 (scientific
funding data).

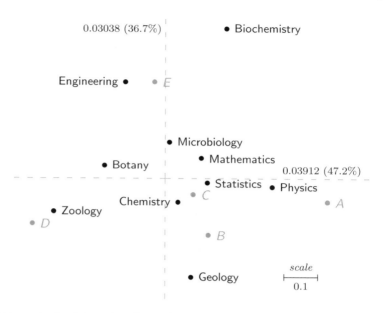

Exhibit 10.3:
Symmetric map of Table 10.1 (scientific funding data).

value of the principal inertias along the axes, or their square roots which are the canonical correlations along each axis, namely $\sqrt{0.039117} = 0.198$ and $\sqrt{0.030381} = 0.174$, respectively. The level of row–column association can be judged graphically only in an asymmetric map such as Exhibit 10.2 (compare again the different levels of association illustrated in Exhibit 4.2).

Dimensional interpretation of maps

Whether the joint map is produced using asymmetric or symmetric scaling, the *dimensional* style of interpretation remains universally valid. This involves interpreting one axis at a time, as we did above and as is customary in factor analysis, using the relative positions of one set of points — the "variables" of the table — to give a descriptive name to the axis. For example, we used the funding category points to give a descriptive name to the axes and then interpreted the discipline points with respect to the axes. All statements in such an interpretation are relative and it is not possible to judge the absolute difference in funding profiles between the disciplines unless we refer to the original data. Putting this another way, symmetric maps similar to Exhibit 10.3 could be obtained for other data sets where there are much larger (or smaller) levels of association between the funding profiles of the disciplines.

Data set 6: Abundances of marine species in sea-bed samples

CA is used extensively to analyse ecological data, and the second example represents a typical data set in marine biology. The data, given partially in Exhibit 10.4, are the counts of 92 marine species identified in 13 samples from the sea-bed in the North Sea. Most of the samples are taken close to an oil-drilling platform where there is some pollution of the sea-bed, while two samples, regarded as reference samples and assumed unpolluted, are taken far from the drilling activities. These data, and biological data of this kind in

SPECIES	STATIONS (SAMPLES)												
	S4	S8	S9	S12	S13	S14	S15	S18	S19	S23	S24	R40	R42
Myri.ocul.	193	79	150	72	141	302	114	136	267	271	992	5	12
Chae.seto.	34	4	247	19	52	250	331	12	125	37	12	8	3
Amph.falc.	49	58	66	47	78	92	113	38	96	76	37	0	5
Myse.bide.	30	11	36	65	35	37	21	3	20	156	12	58	43
⋮	⋮	⋮	⋮	⋮	⋮	⋮	⋮	⋮	⋮	⋮	⋮	⋮	⋮
Eucl.sp.	0	0	0	0	1	0	0	1	1	0	0	0	0
Scal.infl.	0	1	0	0	0	1	0	0	0	0	0	0	1
Eumi.ocke.	0	0	1	0	0	1	1	0	0	0	0	0	0
Modi.modi.	0	0	0	1	1	0	0	1	0	0	0	0	0

Exhibit 10.4:
Frequencies of 92 marine species in 13 samples (the last two are reference samples); the species (rows) have been ordered in descending order of total abundance; hence four most abundant and four least abundant are shown here.

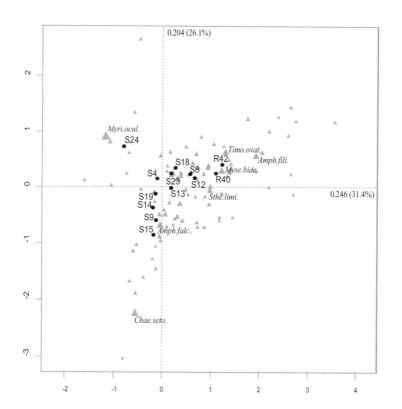

Exhibit 10.5:
Asymmetric CA map, with stations in principal coordinates and the species in standard coordinates. The species symbols have size proportional to the species abundance (mass) — some important species in the analysis are labelled with the first letter of the label being close to its corresponding triangular symbol. Inertia explained in map: 57.5%.

general, are characterized by high variability, which can already be seen by simple inspection of the small part of the data given here. The total inertia of this table is 0.7826, much higher than in the previous examples, so we can expect the profiles to be more spread out relative to the vertices. Notice that in this example the χ^2-test is not applicable, since the data do not constitute a true contingency table — each individual count is not independent of the others, since the marine organisms often occur in groups at a sampling point.

Asymmetric CA map of species abundance data

Exhibit 10.5 shows the asymmetric map of the sample (column) profiles and species (row) vertices. Since there are 92 species points, it is impossible to label each point so we have labelled only the points which have a high contribution to the map; these are generally the most abundant ones. (The topic of how to measure this contribution is described in the Chapter 11, for the moment let us simply report that 10 out of the 92 species contribute over 85% to the construction of this map, the other 82 could effectively be removed without the map changing very much.) The stations form a curve from bottom left (actually, the most polluted stations) to top right (the least polluted), with the reference stations far from the drilling area at upper right. An exception is station 24, which separates out notably from the others, mainly because of the very high abundance of species *Myri.ocul.* (*Myriochele oculata*) which can be seen in the first row of Exhibit 10.4. The most abundant species are labelled and it is mainly these that determine the map. Notice that the asymmetric map does well in this example because the inertia is so high, which is typical of ecological data where there is high variability between the samples. The next example is the complete opposite!

Exhibit 10.6:
Letter counts in 12 samples of texts from books by six different authors, showing data for 9 out 26 letters.

BOOKS	a	b	c	d	e	$\cdots$	w	x	y	z	*Sum*
TD-Buck	550	116	147	374	1015	$\cdots$	155	5	150	3	*7144*
EW-Buck	557	129	128	343	996	$\cdots$	187	10	184	4	*7479*
Dr-Mich	515	109	172	311	827	$\cdots$	156	14	137	5	*6669*
As-Mich	554	108	206	243	797	$\cdots$	149	2	80	6	*6510*
LW-Clar	590	112	181	265	940	$\cdots$	146	13	162	10	*7100*
PF-Clar	592	151	251	238	985	$\cdots$	106	15	142	20	*7505*
FA-Hemi	589	72	129	339	866	$\cdots$	225	1	155	2	*6877*
Is-Hemi	576	120	136	404	873	$\cdots$	250	3	104	5	*6924*
SF7-Faul	541	109	136	228	763	$\cdots$	160	11	280	1	*6885*
SF6-Faul	517	96	127	356	771	$\cdots$	216	12	171	5	*6971*
Pe3-Holt	557	97	145	354	909	$\cdots$	194	9	140	4	*6650*
Pe2-Holt	541	93	149	390	887	$\cdots$	218	2	127	2	*6933*

Abbreviations:
TD (Three Daughters), EW (East Wind) -Buck (Pearl S. Buck)
Dr (Drifters), As (Asia) -Mich (James Michener)
LW (Lost World), PF (Profiles of Future) -Clar (Arthur C. Clarke)
FA (Farewell to Arms), Is (Islands) -Hemi (Ernest Hemingway)
SF7 and SF6 (Sound and Fury, chapters 7 and 6) -Faul (William Faulkner)
Pen3 and Pen2 (Bride of Pendorric, chapters 3 and 2) -Holt (Victoria Holt)

Data set 7: Frequencies of letters in books by six authors

This surprising example is a data set provided in the **ca** package of the R program (see Computational Appendix, pages 222–223). The data form a 12×26 matrix with the rows representing 12 texts which form six pairs, each pair by the same author (Exhibit 10.6 shows a part of the matrix).

The columns are the 26 letters of the alphabet, *a* to *z*. The data are the counts of these letters in a sample of text from each of the books. There are approximately 6500-7500 letter counts for each book or chapter.

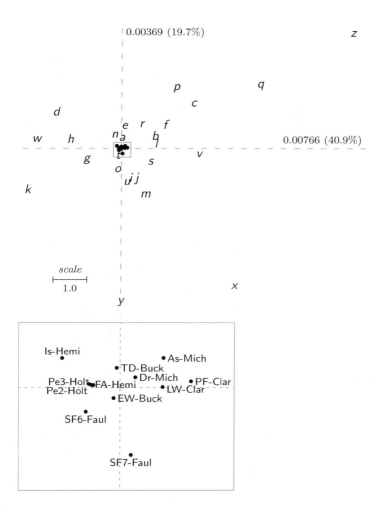

Exhibit 10.7:
Asymmetric CA map of the author data of Table 10.6, with row points (texts) in principal coordinates. The very low inertia in the table is seen in the closeness of the row profiles to the centroid. A "blow-up" of the rectangle at the centre of the map shows the relative positions of the row profiles.

This data set has one of the lowest total inertias I have seen in my experience with CA: the total inertia is 0.01873, which means that the data are very close to the expected values calculated from the marginal frequencies; i.e., the profiles are almost identical. The asymmetric map of these data is shown in Exhibit 10.8, showing the letters in their vertex positions and the 12 texts as a tiny blob of points around the origin, showing how little variation there is between the texts in terms of letter distributions, which is what one would expect. If one expands the tiny blob of points, it is surprising to see how

One of the lowest inertias, but with a significant structure

much structure there is within such tiny variation. Each pair of texts by the same author lies in the same vicinity, and the result is highly significant from a statistical viewpoint (we discuss the permutation test for testing this in Chapter 25).

Preserving a
unit aspect ratio in
maps

An important final remark concerns the physical plotting of two-dimensional correspondence analysis maps. Since distances in the map are of central interest, it is clear that a unit on the horizontal axis of a plot should be equal to a unit on the vertical axis. Even though this requirement seems obvious, it is commonly overlooked in many software packages and spreadsheet programs that produce scatterplots of points with different scales on the axes. For example, the points might in reality have little variation on the vertical second axis, but the map is printed in a pre-defined rectangle which then exaggerates the second axis. We say that the *aspect ratio* of the map, that is the ratio of one unit length horizontally to one unit vertically, should be equal to 1. A few options for producing good quality maps are discussed at the end of the Computational Appendix.

SUMMARY:
Three More
Examples

1. When applicable, it is useful to test a contingency table for significant association, using the χ^2 test. However, statistical significance is not a crucial requirement for justifying an inspection of the maps. CA should be regarded as a way of re-expressing the data in pictorial form for ease of interpretation — with this objective any table of data is worth looking at.

2. In both asymmetric and symmetric maps the dimensional style of interpretation is valid. This applies to one axis at a time and consists of using the relative positions of one set of points on a principal axis to give the dimension a conceptual name, and then separately interpreting the relative positions of the other set of points along this named dimension.

3. The asymmetric map functions well when total inertia is high, but it is problematic when total inertia is small because the profile points in principal coordinates are too close to the origin for easy labelling.

4. It is important to have plotting facilities which preserve the *aspect ratio* of the display. A unit on the horizontal axis must be as close as possible to a unit on the vertical axis of the map; otherwise distances will be distorted if the scales are different.

Contributions to Inertia

The total inertia of a cross-tabulation is a measure of how much variation there is in the table. We have seen how this inertia is decomposed along principal axes and also how it is decomposed amongst the rows or amongst the columns. The inertia can be further broken down into row and column components along individual principal axes. The investigation of these components of inertia (analogous to an analysis of variance) plays an important supporting role in the interpretation of CA. They provide diagnostics which allow the user to identify which points are the major contributors to a principal axis and to gauge how well individual points are displayed.

Contents

In Chapter 4, Equation (4.7), we saw that the total inertia can be interpreted geometrically as the weighted average of squared χ^2-distances between the profiles and their average profile, and is identical for row profiles and for column profiles. If there are only small differences between the profiles and their average, then the inertia is close to zero; i.e., there is low variation (see Exhibit 4.2, top left display). At the other extreme, if each profile is highly concentrated in a few categories, and in different categories from profile to profile, then the inertia is high (Exhibit 4.2, lower right display). The inertia is a measure of how spread out the profiles are in the profile space.

Total inertia measures overall variation of the profiles

Row and column inertias There are various ways that the inertia can be decomposed into the sum of positive components, and this provides a numerical "analysis of inertia" which is helpful in interpreting the results of CA. According to Equation (4.6), each row makes a positive contribution to the inertia in the form of its mass times squared distance to the row centroid — we call this the *row inertia*. The same applies to the columns, leading to *column inertias*. The actual values of these parts of inertia are numbers that are inconvenient to interpret and it is easier to judge them relative to the total inertia, expressed as proportions, percentages or more conveniently as *permills* (i.e. thousandths, denoted by ‰). The following table gives the row and column inertias for the scientific funding data of Exhibit 10.1, first in their "raw" form, and then in their relative form expressed as permills (Exhibit 11.1). Permills are used extensively throughout the numerical results of our R implementation of CA (see Computational Appendix), because they enable reporting three significant digits without using a decimal point, improving legibility of the results.

Exhibit 11.1:
Row and column contributions to inertia, in raw amounts which sum up to the total inertia, or expressed relatively as permills (‰) which add up to 1000.

ROWS	*Inertia*	*‰ inertia*	COLUMNS	*Inertia*	*‰ inertia*
Geology	0.01135	137	A	0.01551	187
Biochemistry	0.00990	119	B	0.00911	110
Chemistry	0.00172	21	C	0.00778	94
Zoology	0.01909	230	D	0.02877	347
Physics	0.01621	196	E	0.02171	262
Engineering	0.01256	152			
Microbiology	0.00083	10			
Botany	0.00552	67			
Statistics	0.00102	12			
Mathematics	0.00466	56			
Total	*0.08288*	*1000*	*Total*	*0.08288*	*1000*

Large and small contributions From the "‰ *inertia*" columns in Exhibit 11.1 we can see at a glance that the major contributors to inertia are the rows Zoology, Physics, Engineering, Geology and Biochemistry, in that order, while for the columns the major contributors are categories *D* and *E*. As a general guideline for deciding which contributions are large and which are small, we use the average as a threshold. For example, there are 10 rows, so on a permill scale this would be 100 on average per row; hence we regard rows with contributions higher than 100‰ as major contributors. On the other hand, there are five columns, which give an average of 200‰, so the two columns *D* and *E* are the major contributors.

Cell contributions to inertia A finer look at the inertia contributions can be made by looking at each individual cell's contribution. As described in Chapter 4, each cell of the table contributes a positive amount to the total inertia, which can again be expressed on a permill scale — see Exhibit 11.2. Here we can see specific cells

SCIENTIFIC AREAS	*FUNDING CATEGORIES*					
	A	*B*	*C*	*D*	*E*	*Sum*
Geology	0	32	16	0	89	*137*
Biochemistry	0	23	4	44	48	*119*
Chemistry	3	12	1	0	5	*21*
Zoology	9	15	11	189	8	*230*
Physics	106	11	2	74	3	*196*
Engineering	1	11	38	1	102	*152*
Microbiology	2	0	0	3	5	*10*
Botany	51	4	0	10	2	*67*
Statistics	10	0	0	2	0	*12*
Mathematics	5	3	22	26	0	*56*
Sum	*187*	*110*	*94*	*347*	*262*	*1000*

Exhibit 11.2:
Cell contributions to inertia, expressed as permills; the row and column sums of this table are identical to the row and column inertias in permills given above in Exhibit 11.1.

such as [Zoology,D] and [Physics,A], that are contributing highly to the inertia — just these two cells together account for almost 30% of the table's total inertia ($189 + 106 = 295$‰, i.e., 0.295 of the total inertia, or 29.5%). The cell contributions to inertia are sometimes called chi-square contributions because they are identical to the relative contributions of each cell to the χ^2 statistic. Adding the rows or columns of this table gives the same permill contributions of Exhibit 11.1.

The other major decomposition of inertia is with respect to, or along, principal axes. On page 74 we gave the first two principal inertias for this 10×5 table, which has four dimensions. Exhibit 11.3 gives all the principal inertias, their precentages and a bar chart (this type of bar chart is often called a *scree plot*). We have seen that the principal inertias have an interpretation in their own right, for example as squared canonical correlations (see Chapter 8, page 61), but we mainly interpret their values relative to the total, usually expressed as percentages rather than permills in this case.

Decomposition along principal axes

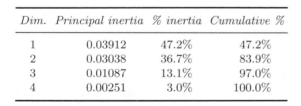 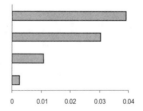

Dim.	Principal inertia	% inertia	Cumulative %
1	0.03912	47.2%	47.2%
2	0.03038	36.7%	83.9%
3	0.01087	13.1%	97.0%
4	0.00251	3.0%	100.0%

0   0.01   0.02   0.03   0.04

Exhibit 11.3:
Principal inertias, percentages and cumulative percentages for all dimensions of the scientific-funding data, and a scree plot.

Each principal inertia is itself an inertia, calculated for the projections of the row profiles (or column profiles) onto a principal axis. For example, the 10 row profiles of the scientific funding data lie in a *full space* of dimensionality 4, one less than the number of columns. Their weighted sum of squared distances to the row centroid is equal to the total inertia, with value 0.08288. The first

Components of each principal inertia

ROWS	*Inertia*	*‰ inertia*	COLUMNS	*Inertia*	*‰ inertia*
Geology	0.00062	16	*A*	0.00890	228
Biochemistry	0.00118	30	*B*	0.00260	67
Chemistry	0.00023	6	*C*	0.00265	68
Zoology	0.01616	413	*D*	0.02471	632
Physics	0.01426	365	*E*	0.00025	6
Engineering	0.00153	39			
Microbiology	0.00001	0			
Botany	0.00345	88			
Statistics	0.00057	14			
Mathematics	0.00112	29			
Total	*0.03912*	*1000*	*Total*	*0.03912*	*1000*

SCIENTIFIC AREAS	*Axis 1*	*Axis 2*	*Axis 3*	*Axis 4*	*Total*
Geology	0.00062	0.00978	0.00082	0.00013	*0.01135*
Biochemistry	0.00118	0.00754	0.00084	0.00034	*0.00990*
Chemistry	0.00023	0.00088	0.00029	0.00032	*0.00172*
Zoology	0.01616	0.00158	0.00063	0.00073	*0.01909*
Physics	0.01426	0.00010	0.00169	0.00016	*0.01621*
Engineering	0.00153	0.00941	0.00127	0.00036	*0.01256*
Microbiology	0.00001	0.00056	0.00008	0.00019	*0.00083*
Botany	0.00345	0.00016	0.00180	0.00011	*0.00552*
Statistics	0.00057	0.00001	0.00042	0.00003	*0.00102*
Mathematics	0.00112	0.00037	0.00302	0.00015	*0.00466*
Total	*0.03912*	*0.03038*	*0.01087*	*0.00251*	*0.08288*

principal axis is the straight line that comes closest to the profile points in the sense of least squares. This axis passes through the row centroid, which is at the *origin*, or zero point, of the display. Suppose that all the row profiles are projected onto this axis. The first principal inertia is then the weighted sum of squared distances from these projections to the centroid. Hence the first principal inertia, equal to 0.03912, is the inertia of the set of projected points on the one-dimensional principal axis. Using the principal coordinates on the axis we obtain the row and column components of the first principal inertia, in Exhibit 11.4. This shows that category *D* is the dominant contributor to the first axis, followed by *A*, while the other categories contribute very little. As for the rows, Zoology (highly associated with *D*) and Physics (highly associated with *A*) contribute almost 78% of the inertia on the first axis.

*Complete
decomposition of
inertia over profiles
and principal axes*

We can repeat the above for all the principal axes, and Exhibit 11.5 shows the raw components of inertia of the rows for all four axes, (a similar table can be constructed for the columns). Just as the raw inertias in Exhibit 11.4 have been expressed in permills relative to the first principal inertia, we could do the same for axes 2 to 4 as well. For example, the major row contributions to axis 2 are Geology, Engineering and Biochemistry. Inspecting these contributions of each row point (and, similarly, each column point) to the principal axes gives numerical support to our interpretation of the map.

Whereas the column sums of Exhibit 11.5 give the principal inertias on respective axes, the row sums give the inertias of the profiles (hence these row sums are the same as the first column of Exhibit 11.1). We can also express these components relative to the row inertias, again either as proportions, percentages or permills. These will tell us how well each row is explained by each principal axis. This is a mini-version of the way we interpreted the principal inertias, which quantified the percentage of the total inertia that was contained on each axis — here we do the same for each point separately. Exhibit 11.6 gives these relative amounts in permills, so that each row now adds up to 1000. For example, Geology is mostly explained by axis 2, whereas Physics

SCIENTIFIC AREAS	*PRINCIPAL AXES*				
	Axis 1	*Axis 2*	*Axis 3*	*Axis 4*	*Total*
Geology	55	861	72	11	*1000*
Biochemistry	119	762	85	35	*1000*
Chemistry	134	510	170	186	*1000*
Zoology	846	83	33	38	*1000*
Physics	880	6	104	10	*1000*
Engineering	121	749	101	28	*1000*
Microbiology	9	671	96	224	*1000*
Botany	625	29	326	20	*1000*
Statistics	554	7	410	30	*1000*
Mathematics	240	79	649	33	*1000*
Average	*472*	*367*	*131*	*30*	*1000*

Exhibit 11.6: *Relative contributions (in ‰) of each principal axis to the inertia of individual points; the last row shows the same calculation for the principal inertias (cf. Exhibit 11.3), which can be regarded as average relative contributions.*

mostly by axis 1. Mathematics, on the other hand, is not well explained by axis 1 or axis 2; in fact, its inertia is mostly in the third dimension.

Exhibit 11.7 illustrates the decomposition of inertia and introduces some notation at the same time. The point $\mathbf{a}_i$ is a general profile point in multidimensional space, for example the i-th row profile, with mass r_i, at a distance of d_i from the average row profile $\mathbf{c}$. Hence, using formula (4.6), the total inertia is equal to $\sum_i r_i d_i^2$. A general principal axis k is shown and the point's principal coordinate on this axis is denoted by f_{ik}. Thus the inertia along this axis (i.e., the k-th principal inertia) is $\sum_i r_i f_{ik}^2$, usually denoted by λ_k. Hence the contribution of each point i to the principal inertia of axis k is $r_i f_{ik}^2$ relative to λ_k (these proportions are given in permills for axis 1 in Exhibit 11.4). Exhibit 11.5 is actually the table of values $r_i f_{ik}^2$ for the 10 rows and 4 principal axes of the scientific funding data, with column sums equal to the λ_k's and the

Algebra of inertia decomposition

Exhibit 11.7:
*Example of a profile
point in
multidimensional
space at a
χ^2-distance d_i from
the centroid,
projecting onto the
k-th principal axis
at the principal
coordinate f_{ik}.*

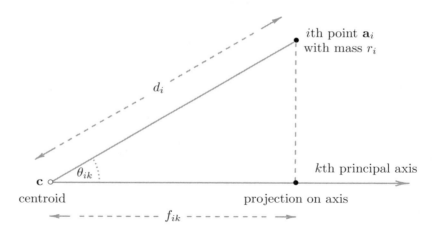

row sums equal to the row inertias $r_i d_i^2$. Thanks to Pythagoras' theorem, we have $d_i^2 = \sum_k f_{ik}^2$, which is why the rows of Exhibit 11.5 sum up to the row inertias:

$$\sum_k r_i f_{ik}^2 = r_i d_i^2$$

Hence, the contribution of axis k to the inertia of point i is $r_i f_{ik}^2$ relative to $r_i d_i^2$ (these proportions are given in Exhibit 11.6).

*Relative
contributions as
squared angle
cosines*

There is an alternative geometric interpretation of the relative contributions in Exhibit 11.6. Since the proportion of inertia of point i explained by axis k is $r_i f_{ik}^2 / r_i d_i^2 = (f_{ik}/d_i)^2$, it is clear from Exhibit 11.7 that this is the square of the angle cosine between the point and the axis. Suppose this angle is denoted by θ_{ik}, then the relative contribution is $\cos^2(\theta_{ik})$: for example, axis 1 has a relative contribution of 0.880 to the point Physics; hence $\cos^2(\theta_{51}) = 0.880$, from which we can evaluate $\cos(\theta_{51}) = 0.938$ and the angle $\theta_{51} = 20°$. This shows that the point Physics, which is mostly explained by axis 1, is close to axis 1, subtending a small angle of 20° with it. A point like Geology, with a relative contribution of 0.055, subtends a large angle of $\theta_{11} = 76°$ with axis 1, so is not at all close to this axis but lying along different dimensions of the space (in fact, mostly along axis 2 as we can see by the high relative contribution of 0.861).

*Relative
contributions as
squared
correlations*

There is a further interpretation of the relative contributions: angles between vectors can be interpreted as correlation coefficients; hence the relative contributions are also squared correlations. We can thus say that Physics has a high correlation of $\sqrt{0.880} = 0.938$ with axis 1, whereas Geology has a low correlation of $\sqrt{0.055} = 0.234$. If the correlation is 1, the profile point lies on the principal axis, and if the correlation is 0 the profile is perpendicular to the principal axis (angle of 90°).

SCIENTIFIC AREAS	Quality	FUNDING CATEGORIES	Quality
Geology	916	A	587
Biochemistry	881	B	816
Chemistry	644	C	465
Zoology	929	D	968
Physics	886	E	990
Engineering	870		
Microbiology	680		
Botany	654		
Statistics	561		
Mathematics	319		
Overall	*839*	*Overall*	*839*

Exhibit 11.8: *Quality of display (in permills) of individual row profile points in two dimensions; only Mathematics has less than 50% of its inertia explained.*

Thanks to Pythagoras' theorem, the squared cosines of the angles between a point and each of a set of axes can be added together to give squared cosines between the point and the subspace generated by those axes. For example, the angle between a row profile and the principal plane can be computed from the sum of the relative contributions along the first two principal axes. Exhibit 11.7 gives the sum of the first two columns of Exhibit 11.6, and these are interpreted as measures of *quality* of individual points in the two-dimensional maps of Chapter 10, just as the sum of the first two percentages of inertia is interpreted as a measure of overall (or average) quality of display. Here we can see which points are well represented in the two-dimensional display and which are not. Putting this another way, since 83.9% of the inertia is explained in the two-dimensional map, 16.1% of the inertia is not explained. Some profiles will not be accurately represented because they lie more along the third and fourth axes than along the first two. Thus **Mathematics** is poorly displayed, with over two-thirds of its inertia lying off the plane. In Exhibits 10.2 and 10.3 **Mathematics** looks quite similar to the profile of **Statistics**, but this projected position is not an accurate reflection of its true position.

Quality of display in a subspace

This section is mainly aimed at readers with a knowledge of factor analysis — several entities in CA have direct analogues with those in factor analysis.

Analogy with factor analysis

- The analogue of a *factor loading* is the angle cosine between a point and an axis, i.e., the square root of the squared correlation with the sign of the point's coordinate. For example, from Exhibits 11.1 and 11.4, the squared correlations of the categories *A* to *E* are:

$$A: \frac{0.00890}{0.01551} = 0.574 \qquad B: \frac{0.00260}{0.00911} = 0.286 \qquad C: \frac{0.00265}{0.00778} = 0.341$$

$$D: \frac{0.02471}{0.02877} = 0.859 \qquad E: \frac{0.00025}{0.02171} = 0.012$$

Using the signs of the column coordinates in Exhibit 10.3, the CA "factor loadings" would be the signed square roots:

$$A: 0.758 \quad B: 0.535 \quad C: 0.584 \quad D: -0.927 \quad E: -0.108$$

- The analogue of a *communality* is the quality measure on a scale of 0 to 1. For example, in the two-dimensional solution, the CA "communalities" of the five column categories is given by the last column of Exhibit 11.8 on the original scale: 0.587, 0.816, 0.465, 0.968 and 0.990, respectively.

- The analogue of a *specificity* is 1 minus the quality measure on a scale of 0 to 1. For example, in the two-dimensional solution, th e CA "specificities" of the five column categories are 0.413, 0.184, 0.535, 0.032 and 0.010, respectively.

SUMMARY:
Contributions to
Inertia

1. The (total) inertia of a table quantifies how much variation is present in the set of row profiles or in the set of column profiles.

2. Each row and each column makes a contribution to the total inertia, called a *row inertia* and a *column inertia*, respectively.

3. CA is performed with the objective of accounting for a maximum amount of inertia along the first axis. The second axis accounts for a maximum of the remaining inertia, and so on. Thus the total inertia is also decomposed into components along principal axes, i.e., the principal inertias.

4. The principal inertias are themselves decomposed over the rows and the columns. These *inertia contributions* are more readily expressed in relative amounts, and there are two possibilities:

    (a) express each contribution to the k-th axis relative to the corresponding principal inertia.

    (b) express each contribution to the k-th axis relative to the corresponding point's inertia.

5. Possibility (a) allows diagnosing which points have played a major part in determining the orientation of the principal axes. These contributions facilitate the interpretation of each principal axis.

6. Possibility (b) allows diagnosing the position of each point and whether a point is well represented in the map, in which case the point is interpreted with confidence, or poorly represented, in which case its position is interpreted with more caution. These quantities are squared cosines between the points and the principal axes, also interpreted as squared correlations.

7. The sum of squared correlations for a point in a low-dimensional solution space gives a measure of *quality* of representation of the point in that space.

8. The correlations of the points with the axes are the analogues of factor loadings, and the qualities are analogues of communalities.

Supplementary Points

It frequently happens that there are additional rows and columns of data that are not the primary data of interest but that are useful in interpreting features discovered in the primary data. Any additional row (or column) of a data matrix can be positioned on an existing map, as long as the profile of this row (or column) is meaningfully comparable to the existing row (or column) profiles which have determined the map. These additional rows or columns that are added to the map afterwards are called *supplementary points*.

Contents

Up to now all rows and all columns of a particular table of data have been used to determine the principal axes and hence the map — we say that all rows and columns are *active* in the analysis. One can think of each active point having a different force of attraction for the principal axes, where this force depends on the position of the point as well as its mass. Profiles further from the average have more "leverage" in orienting the map towards them, and higher mass profiles have a greater "pull" on the map.

Active points

There are situations, however, when we wish to suppress some points from the actual computation of the solution while still being able to inspect their projections onto the map which best fits the active points. The simplest way to think of such points is that they have a position but no mass at all, so that their contribution to the inertia is zero and they have no influence on the principal axes. Such zero mass points are called *supplementary points*, sometimes also called *passive points* to distinguish them from the active points which have positive mass. There are three common situations when supplementary rows or columns can be useful, and we now illustrate each of these in the context of the scientific funding data set of the previous chapters. Exhibit 12.1 shows

Definition of a supplementary point

Exhibit 12.1:
*Frequencies of
funding categories
for 796 researchers
(Exhibit 10.1), with
additional column Y
for a new category
of "promising young
researchers", an
additional row for
researchers at
museums, and a row
of cumulated
frequencies for
Statistics and
Mathematics,
labelled Math
Sciences.*

SCIENTIFIC AREAS	*FUNDING CATEGORIES*					
	A	*B*	*C*	*D*	*E*	*Y*
Geology	3	19	39	14	10	*0*
Biochemistry	1	2	13	1	12	*1*
Chemistry	6	25	49	21	29	*0*
Zoology	3	15	41	35	26	*0*
Physics	10	22	47	9	26	*1*
Engineering	3	11	25	15	34	*1*
Microbiology	1	6	14	5	11	*1*
Botany	0	12	34	17	23	*1*
Statistics	2	5	11	4	7	*0*
Mathematics	2	11	37	8	20	*1*
Museums	*4*	*12*	*11*	*19*	*7*	
Math Sciences	*4*	*16*	*48*	*12*	*27*	

an expanded version of that data set where we have added:

1. an additional column, labelled Y, which is a special category of funding for young researchers, a category which had just been introduced into the funding system;

2. an additional row, labelled *Museums*, containing the frequencies of researchers working at museums (as opposed to universities, in the rest of the table);

3. another row, labelled *Math Sciences*, which is the sum of the rows **Statistics** and **Mathematics**.

*First case — a
point inherently
different from the
rest*

The study from which these data are derived was primarily aimed at university researchers. Researchers from museums, however, were similarly graded and sponsored by the same funding organization, hence the frequencies of 53 museum researchers in the five funding categories. While it is necessary to consider the museum researchers separately from those at universities, it is still of interest to visualize the profile of museum researchers in the "space" of the university researchers, which can be done by declaring the row *Museums* to be a supplementary point. Its profile does not participate in the determination of the principal axes, but its profile can be projected onto the map. Exhibit 12.2 shows the symmetric map, as in Exhibit 10.2, with the additional point *Museums* in the lower left-hand side of the map. This point has no contribution to the principal inertia, but we can still look at the contributions of the axes to the point (i.e., the relative contributions or squared cosines or squared correlations). It turns out that this point is quite well displayed in the map, with over 50% of its inertia explained by it. Its position indicates that relatively few of the museum researchers have their applications rejected, while those that do receive funding tend towards the lower categories. Various

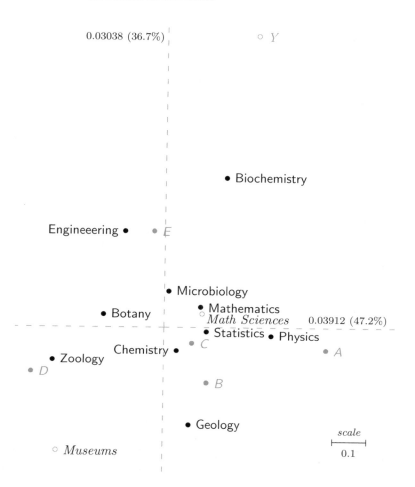

0.03038 (36.7%)

Exhibit 12.2:
*Symmetric map of
the data of Exhibit
12.1 (cf. Exhibit
10.2), showing in
addition the profile
positions of the
supplementary
column Y and
supplementary rows
Museums and Math
Sciences.*

types of supplementary information may be added to an active data set. Such information may be part of the same study, as in the case of *Museums* above, or it may come from separate but similar studies. For example, a similar table of frequencies may be available from a previous classification of scientific researchers, and may be added as a set of supplementary rows in order to trace the evolution of each discipline's funding position over time. Another example is when some target profiles for the disciplines are specified and we want to judge how far away their actual positions are from the targets. This concept of an "ideal point" is frequently used in product positioning studies in marketing research.

Because the additional *Y* category had only just been introduced into the funding system, very few researchers were allocated to that category, in fact only six researchers and each one in a different discipline. This means that the

*Second case — an
outlier of low mass*

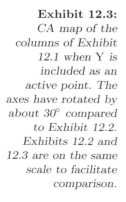

Exhibit 12.3:
CA map of the columns of Exhibit 12.1 when Y is included as an active point. The axes have rotated by about 30° compared to Exhibit 12.2. Exhibits 12.2 and 12.3 are on the same scale to facilitate comparison.

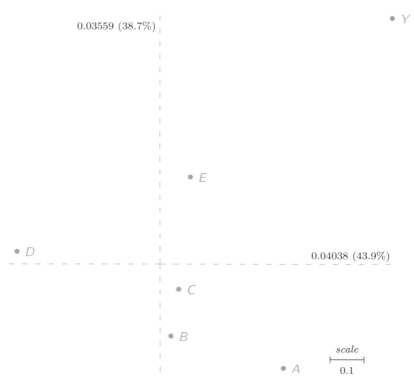

profile of this column is quite unusual: six of the profile values have the value $\frac{1}{6} = 0.167$ and the others are 0. No other column profile has the slightest resemblance to this one, so it is to be expected that it has a very unusual position in the multidimensional space. As we see in Exhibit 12.2, this point is an *outlier* and if it were to be included as an active point in the analysis, it may contribute greatly to the map. This would not be a satisfactory situation since only six people are contained in the column Y — hence, apart from the substantive reason for making it supplementary, there is also a technical one. In this particular case, if we do include Y as an active point, its mass is less than 1% of the columns, but the total inertia of the table increases from 0.0829 to 0.0920, an increase of 11%. In addition, the map changes substantially, as can be seen in Exhibit 12.3 — there appears to be an approximate 30° rotation in the solution compared to the previous solution; hence the inclusion of Y has swung the axes around. We should be on the lookout for such outlying points with low mass that contribute highly to the inertia of the solution. In some extreme cases outliers can start to dominate a map so much that the more interesting contrasts between the more frequently occurring categories are completely masked. By declaring outliers supplementary their positions can still be approximately visualized without influencing the solution space. Another way of dealing with rows or columns of low mass is to combine them with other rows or columns in a way which conforms to the data context: if we had an additional discipline, for example "Computer Science" with very few researchers in this category and a possibly strange profile as a result, we could combine them with an allied field, say Mathematics or Engineering. Having said this, it is a fact that outliers of low mass are often not such a serious problem in CA, since influence is measured by mass times squared distance and the low mass decreases the influence. The real problem is the fact that they lie so far from the other points — we return to this subject in Chapter 13 when we discuss alternative scalings for the map.

Supplementary points can be used to display a group of categories or to display subdivisions of a category. For example, the additional row *Math Sciences* in Exhibit 12.1 is the sum of the frequencies for **Mathematics** and **Statistics**, two disciplines which are frequently grouped together. The profile of this new row is the centroid of the two component rows, which are weighted by their respective masses. Since there are 78 and 29 researchers in Mathematics and Statistics respectively, the profile of Math Sciences would thus be:

*Third case —
displaying groups
or partitions of
points*

$$\textit{Math Sciences} \text{ profile} = \frac{78}{107} \times \textsf{Mathematics profile} + \frac{29}{107} \times \textsf{Statistics profile}$$

so that the *Math Sciences* profile would be more like the **Mathematics** profile than the **Statistics** one. Geometrically, this means that the point representing the profile of *Math Sciences* is on a line between the **Mathematics** and **Statistics** points, but closer to **Mathematics** (cf. Exhibit 3.5 on page 23). In order to display the point *Math Sciences*, as in Exhibit 12.2, the new row is declared to be a supplementary point. We would not make this point active along with its two component rows, since this would mean that the 107 researchers in these two disciplines would be counted twice in the analysis. In the same way, subdivisions of categories may be displayed on existing CA maps. Suppose that data were available for a breakdown of **Engineering** into its different branches, for example mechanical, civil, electrical, etc. Then, to investigate whether the profiles of these subgroups lie in the same general region, these additional rows of frequencies can simply be declared supplementary. The result described above still applies: in the map the active **Engineering** point would be at the centroid of all the points representing its different branches.

In the above we have described supplementary points as additional profile points that are projected onto a previously computed map. An alternative way of obtaining their positions is to position them relative to the set of vertex points in an asymmetric map. For example, in Chapter 3 it was shown that the position of a row profile, say, is a weighted average of the column vertices, where the weights are the profile elements. A supplementary point can be positioned in exactly the same way. Once the principal axes of the row profiles have been determined, we know the coordinate positions of the vertex points representing the columns on each principal axis, i.e., the standard coordinates of the columns. An extra row profile can now be placed on any map

*Computing the
position of a
supplementary
point*

by evaluating the appropriate centroid of the vertices on each principal axis of the map, using the elements of the new profile as weights. For example, to calculate the position of the supplementary point *Museums*:

$$\text{position of } \textit{Museums} = \frac{4}{53} \times \text{vertex } A + \frac{12}{53} \times \text{vertex } B + \cdots \text{etc.}$$

i.e., calculate the weighted average of the standard coordinates of the columns along each principal axis.

Contributions of supplementary points

Since supplementary points have zero mass, they also have zero inertia and make no contribution to the principal inertias. Their relative contributions, which relate to the angles between profiles and axes and do not involve masses, can still be interpreted to diagnose how well they are represented. The relative contributions and qualities in the two-dimensional space of the three supplementary points described above are as follows:

SUPPLEMENTARY POINTS	*Relative contributions*		*Quality*
	Axis 1	*Axis 2*	*in 2-dimns*
Museums	225	331	556
Math Sciences	493	66	559
Y	4	587	641

These quantities describe how well these additional points are being displayed. For example, the supplementary point *Y* subtends an angle whose squared cosine is 0.054 with the first axis and 0.587 with the second axis. Its quality of display in the plane is thus $0.054 + 0.587 = 0.641$, so that 64.1% of its position is contained in the plane, and 35.9% in the remaining dimensions. Or we can say that *Y* is correlated $\sqrt{0.641} = 0.801$ with the plane.

Vertices are supplementary points

We have already encountered supplementary points in the form of the vertex points which we projected onto maps for purposes of interpretation, but whose positions were not taken into account in computing the map itself. This suggests an alternative way of determining the positions of the vertices: firstly, increase the data set by a number of rows, as many as there are columns of data, each of which consists of zeros except for a single 1, where this 1 is in a different column of each row (Exhibit 12.4); and secondly, declare these

Exhibit 12.4:
Supplementary rows which could be added to the table in Exhibit 12.1 — their positions are identical to the column vertex points.

FUNDING CATEGORIES	*A*	*B*	*C*	*D*	*E*
A	1	0	0	0	0
B	0	1	0	0	0
C	0	0	1	0	0
D	0	0	0	1	0
E	0	0	0	0	1

additional rows to be supplementary points. The positions of these supplementary rows are identical to those of the column vertices; in other words their coordinates will be the standard coordinates of the columns.

The example of column Y and the vertex points in Exhibit 12.4 should not be confused with what is called "dummy variable" coding, a subject which we shall treat in detail when we come to multiple correspondence analysis in later chapters. For example, suppose that we had a classification of the scientific areas into "Natural Sciences" (NS) and "Biological Sciences" (BS), the latter group including Biochemistry, Zoology, Microbiology and Botany while the former group contains the rest. A standard way of coding this in CA is as a pair of dummy variables, NS and BS say, zero-one variables with the values $NS = 1$ and $BS = 0$ for Geology (a natural science), for example, and $NS = 0$ and $BS = 1$ for Biochemistry (a biological science), and so on. One might be tempted to add these dummy variables as columns of the table and display them as supplementary points, but this would not be correct. This is not a count variable like the Y variable, which happened to have had 0's and 1's as well; in that case the data were real counts and could have been other integer values. The correct way to display this NS/BS information is as a pair of rows, similar to the way we displayed *Math Sciences* above. That is, sum up the frequencies for the NS rows and add an extra row called NS to the table, and do the same for the BS rows. In this way the NS and BS points will be weighted averages of the points representing the two sets of scientific areas (we shall return to this subject in Chapter 18).

Categorical supplementary variables and dummy variables

Additional information in the form of continuous variables also needs special consideration. Suppose we had some external information about each scientific area, for example, the average impact factor of all papers published in these areas in international journals. This would also be stored as a column of data, and because these are all positive numbers one might be tempted to represent the profile of this column in the standard supplementary point fashion. But remember that it is the profile of the column that is represented, not the original numbers, so the values should be nonnegative and expressing them as proportions of the total should make sense in the context of the study. But what if the data were changes in the average impact factors over a period of time, so that some changes were positive and some negative? Clearly, expressing these changes relative to their sum makes no sense. In this situation the continuous variable can be depicted in the map in a completely different way, using regression analysis. This subject will be treated in more detail in Chapters 13 and 14 and also in Chapter 24 on canonical correspondence analysis, which is a combination of CA and regression — for the moment, we merely alert the reader to the problem.

Continuous supplementary variables

SUMMARY:
Supplementary
Points

1. The rows and columns of a table analysed by CA are called *active* points. These are the points that determine the orientation of the principal axes and thus the construction of low-dimensional maps. The active rows and columns are projected onto the map.

2. *Supplementary* (or *passive*) points are additional rows or columns of a table which have meaningful profiles and which exist in the full spaces of the row and column profiles respectively. They can also be projected onto a low-dimensional map in order to interpret their positions relative to the active points.

3. Since supplementary points have zero mass, all quantities involving the mass, the point inertia and the contribution of the point to an axis, are also zero.

4. The contribution of each principal axis to a supplementary point (i.e., squared cosine or squared correlation) can be computed and allows an assessment of whether a supplementary point lies to a larger or lesser extent in the subspace of the map. For example, the map might explain the supplementary point quite well even though the supplementary point has not determined the solution.

5. Be on the lookout for outliers with low mass — their presence in the analysis might have high influence on the solution. If they do, they should be made supplementary or combined with other rows or columns in a substantively sensible way.

6. A supplementary categorical variable, for example a column, should be used to agglomerate the rows according to its categories and then add these categories as supplementary rows of the table.

7. Care is needed when adding an external continuous variable as a supplementary point: its values have to be nonnegative and its profile must make sense in the context of the example.

Correspondence Analysis Biplots

Up to now we have drawn and interpreted CA maps in two possible ways. In the asymmetric map, for example in the row analysis, the χ^2-distances between row profiles are displayed as accurately as possible, taking into account the masses of each profile, while the column vertices serve as references for the interpretation. In the symmetric map, the rows and the columns are both represented as profiles, thus the χ^2-distances between row profiles and between column profiles are approximated. The *biplot* is an alternative way of interpreting a joint map of row and column points. This approach is based on the scalar products between row vectors and column vectors, which depend on the lengths of the vectors and the angles between them rather than their interpoint distances. In the biplot only one of the profile sets, either the rows or the columns, are represented in principal coordinates. In fact, asymmetric CA maps, with one set in principal coordinates and the other in standard coordinates, are biplots. But there are alternative choices of coordinates for the other set of points serving as the references for the interpretation.

Contents

In Euclidean geometry a *scalar product* between two vectors $\mathbf{x}$ and $\mathbf{y}$ with coordinates x_1, x_2, ... and y_1, y_2, ... is the sum of products of respective elements $x_k y_k$, denoted by $\mathbf{x}^\mathsf{T}\mathbf{y} = \sum_k x_k y_k$ ($^\mathsf{T}$ is the notation for the transpose of a vector or a matrix). Geometrically the scalar product is equal to the product of the lengths of the two vectors, multiplied by the cosine of the

Definition of a scalar product

angle between them:

$$\mathbf{x}^\mathsf{T}\mathbf{y} = \sum_k x_k y_k = \|\mathbf{x}\| \cdot \|\mathbf{y}\| \cdot \cos\theta \qquad (13.1)$$

where $\|\mathbf{x}\|$, for example, denotes the length of the vector $\mathbf{x}$, i.e., the distance between the point $\mathbf{x}$ and the zero point. This result is illustrated in two-dimensional space in Exhibit 13.1 (notice that two vectors in multidimensional space can always be represented in a plane).

Exhibit 13.1:
Example of two points $\mathbf{x}$ and $\mathbf{y}$ whose vectors subtend an angle of θ with respect to an origin (usually the centroid of the cloud of points). The scalar product between the points is the length of the projection of $\mathbf{x}$, say, onto $\mathbf{y}$, multiplied by the length of $\mathbf{y}$.

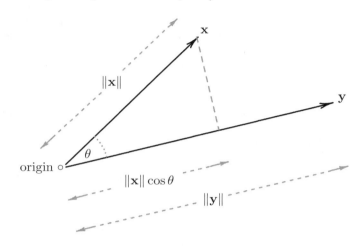

Relationship between scalar product and projection — Another standard geometric result is that the perpendicular projection of a vector $\mathbf{x}$ onto a direction defined by a vector $\mathbf{y}$ has a length equal to the length of $\mathbf{x}$ multiplied by the cosine of the angle between $\mathbf{x}$ and $\mathbf{y}$, i.e., the $\|\mathbf{x}\| \cdot \cos\theta$ part of the definition (13.1). Thus, the scalar product between $\mathbf{x}$ and $\mathbf{y}$ can be thought of as the projected length of $\mathbf{x}$ onto $\mathbf{y}$ multiplied by the length of $\mathbf{y}$ (illustrated in Exhibit 13.1), or, equivalently, as the projected length of $\mathbf{y}$ onto $\mathbf{x}$ multiplied by the length of $\mathbf{x}$. If the length of one of the vectors, say $\mathbf{y}$, is one, then the scalar product is simply the length of the projection of the other vector $\mathbf{x}$ onto $\mathbf{y}$.

For fixed reference vector, scalar products are proportional to projections — If we think of $\mathbf{y}$ as a fixed reference vector, and then imagine several vectors $\mathbf{x}_1, \mathbf{x}_2, \dots$ projecting onto $\mathbf{y}$, then it is clear that

- the scalar products $\mathbf{x}_1^\mathsf{T}\mathbf{y}, \mathbf{x}_2^\mathsf{T}\mathbf{y}, \dots$ have magnitudes proportional to the projections, since they are the projections multiplied by the fixed length of $\mathbf{y}$;

- the sign of a scalar product is positive if the vector $\mathbf{x}$ makes an acute angle ($< 90°$) with $\mathbf{y}$ and it is negative if the angle is obtuse ($> 90°$).

These properties are the basis for the biplot interpretation of CA.

The *biplot* is a low-dimensional display of a rectangular data matrix where the rows and columns are represented by points, with a specific interpretation in terms of scalar products. The idea is to recover the individual elements of the data matrix approximately in these scalar products. As an initial example of a biplot that recovers the data exactly, consider the following 5×4 table, denoted by $\mathbf{T}$:

$$\mathbf{T} = \begin{bmatrix} 8 & 2 & 2 & -6 \\ 5 & 0 & 3 & -4 \\ -2 & -3 & 3 & 1 \\ 2 & 3 & -3 & -1 \\ 4 & 6 & -6 & -2 \end{bmatrix} \tag{13.2}$$

and then compare it to the map in Exhibit 13.2, where we also give the coordinates of each point. (Notice the convention in matrix algebra to denote vectors as columns, so that a vector is transposed if it is written as a row.)

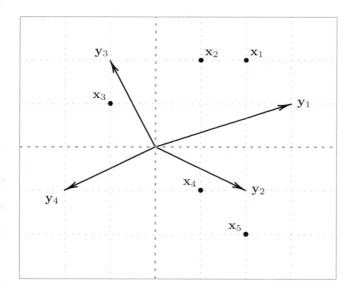

$\mathbf{x}_1 = [\ 2 \quad 2\ ]^\mathsf{T}$

$\mathbf{x}_2 = [\ 1 \quad 2\ ]^\mathsf{T}$

$\mathbf{x}_3 = [\ -1 \quad 1\ ]^\mathsf{T}$

$\mathbf{x}_4 = [\ 1 \quad -1\ ]^\mathsf{T}$

$\mathbf{x}_5 = [\ 2 \quad -2\ ]^\mathsf{T}$

$\mathbf{y}_1 = [\ 3 \quad 1\ ]^\mathsf{T}$

$\mathbf{y}_2 = [\ 2 \quad -1\ ]^\mathsf{T}$

$\mathbf{y}_3 = [\ -1 \quad 2\ ]^\mathsf{T}$

$\mathbf{y}_4 = [\ -2 \quad -1\ ]^\mathsf{T}$

Exhibit 13.2:
Map of five row points $\mathbf{x}_i$ and four column points $\mathbf{y}_j$. The scalar product between the i-th row point and the j-th column point gives the (i,j)-th value t_{ij} of the table in (13.2). The column points are drawn as vectors to encourage the interpretation of the scalar products as projections of the points onto the vectors, multiplied by the respective lengths of the vectors.

For example, the scalar product between $\mathbf{x}_1$ and $\mathbf{y}_1$ is equal to $2 \times 3 + 2 \times 1 = 8$, the first element of $\mathbf{T}$. Just to show that (13.1) can also be used, although with much more trouble, first calculate the respective angles that $\mathbf{x}_1$ and $\mathbf{y}_1$ make with the horizontal axis, using basic trigonometry: $\arctan(2/2) = 45°$ and $\arctan(1/3) = 18.43°$, respectively; hence the angle between $\mathbf{x}_1$ and $\mathbf{y}_1$ is $45 - 18.43 = 26.57°$. Equation (13.1) thus gives the scalar product as:

$$\mathbf{x}_1^\mathsf{T} \mathbf{y}_1 = \|\mathbf{x}_1\| \cdot \|\mathbf{y}_1\| \cdot \cos\theta = \sqrt{8} \cdot \sqrt{10} \cdot \cos(26.57°) = 8.00$$

so this checks. The projection of $\mathbf{x}_1$ onto $\mathbf{y}_1$ is equal to $\sqrt{8}\cos(26.57°) = 2.530$, and the length of $\mathbf{y}_1$ is $\sqrt{10} = 3.162$, the product of which is 8.00.

Some special patterns in biplots

The "bi" in the name biplot comes from the fact that both rows and columns are displayed in a map, not from the bidimensionality of the map — biplots could be of any dimensionality, but the most common case is the planar one. The points in Exhibit 13.2 have been chosen to illustrate some other properties of a biplot:

- $\mathbf{x}_2$ and $\mathbf{y}_2$ are at right angles, so $\mathbf{x}_2$ projects onto the origin; hence the value t_{22} in table $\mathbf{T}$ is 0.

- $\mathbf{x}_2$ and $\mathbf{x}_3$ project onto $\mathbf{y}_3$ at the same point; hence the values t_{23} and t_{33} are equal ($= 3$ in this case).

- $\mathbf{x}_5$ is opposite $\mathbf{x}_3$ with respect to the origin and twice as far away, that is $\mathbf{x}_5 = -2\mathbf{x}_3$; hence the fifth row of table $\mathbf{T}$ is twice the third row, with a change of sign.

- $\mathbf{x}_3$, $\mathbf{x}_4$ and $\mathbf{x}_5$ are on a straight line (this could be any straight line, not necessarily through the origin), so they have a linear relationship, specifically $\mathbf{x}_4 = \frac{1}{3}\mathbf{x}_3 + \frac{2}{3}\mathbf{x}_5$; this weighted average relationship carries over to the corresponding rows of $\mathbf{T}$, for example $t_{41} = \frac{1}{3}t_{31} + \frac{2}{3}t_{51} = \frac{1}{3}(-2) + \frac{2}{3}(4) = 2$.

Rank and dimensionality

In mathematics we would say that the *rank* of the matrix $\mathbf{T}$ in (13.2) is equal to 2, and this is why the table can be perfectly reconstructed in a two-dimensional biplot. In our geometric approach, rank is equivalent to dimensionality.

Biplots give optimal approximations of real data

In real life, a data matrix has higher dimensionality and cannot be reconstructed exactly in a low-dimensional biplot. The idea of the biplot is to find row points $\mathbf{x}_i$ and column points $\mathbf{y}_j$ such that the scalar products between the row and column vectors approximate the corresponding elements of the data matrix as closely as possible. So we can say that the biplot models the data t_{ij} as the sum of a scalar product in some low-dimensional subspace (say K^* dimensions) and a residual "error" term:

$$t_{ij} = \mathbf{x}_i^\mathsf{T}\mathbf{y}_j + e_{ij}$$
$$= \sum_{k=1}^{K^*} x_{ik}y_{jk} + e_{ij} \tag{13.3}$$

This biplot "model" is fitted by minimizing the errors, usually by least squares where $\sum_i \sum_j e_{ij}^2$ is minimized. This looks just like a multiple linear regression equation, except that there are two sets of unknown parameters, the row coordinates $\{x_{ik}\}$ and the column coordinates $\{y_{jk}\}$ — we shall return to the connection with regression analysis in Chapter 14.

The CA model

To understand the link between CA and the biplot, we need to introduce a mathematical formula which expresses the original data n_{ij} in terms of the row and column masses and coordinates. One version of this formula, known as the *reconstitution formula* (see Theoretical Appendix, page 204), is:

$$p_{ij} = r_i c_j \left(1 + \sum_{k=1}^{K} \sqrt{\lambda_k} \phi_{ik} \gamma_{jk}\right) \qquad (13.4)$$

where

- p_{ij} are the relative proportions n_{ij}/n, n being the grand total $\sum_i \sum_j n_{ij}$
- r_i and c_j are the row and column masses
- λ_k is the k-th principal inertia
- ϕ_{ik} and γ_{jk} are row and column standard coordinates, respectively.

In the summation in (13.4) there are as many terms K as there are dimensions in the data matrix, which we have seen to be equal to one less than the number of rows or columns, whichever is smaller. If we map the CA solution in K^* dimensions, where K^* is usually 2, then the fit is optimal since the terms in (13.4) from K^*+1 onwards are minimized — these latter terms thus constitute the "error", or residual.

Equation (13.4) can be slightly re-arranged so that the right-hand side is in the form of a scalar product in a space of dimensionality K^*, plus an error term, as in (13.3):

$$\frac{p_{ij}}{r_i c_j} - 1 = \sum_{k=1}^{K^*} f_{ik} \gamma_{jk} + e_{ij} \qquad (13.5)$$

Biplot of contingency ratios

where $f_{ik} = \sqrt{\lambda_k} \phi_{ik}$, the principal coordinate of the ith row on the kth axis. This shows that the row asymmetric map, which displays row principal coordinates f_{ik} and column principal coordinates γ_{jk}, is an approximate biplot of the values on the left-hand side of (13.5). The ratios $p_{ij}/(r_i c_j)$ of observed proportions to expected ones are called *contingency ratios* — the closer these ratios are to 1, the closer the data are to the independence (or homogeneity) model.

We can also write (13.5) from the row profile point of view as:

$$\left(\frac{p_{ij}}{r_i} - c_j\right)/c_j = \sum_{k=1}^{K^*} f_{ik} \gamma_{jk} + e_{ij} \qquad (13.6)$$

Biplot from row profile point of view

which shows that the row asymmetric map is also a biplot of the deviations of the row profiles from their average, relative to their average (see, for example, Exhibit 10.2). As we have seen, however, the asymmetric map can be quite unsatisfactory when inertia is small, because the row profiles (with coordinates f_{ik}) are concentrated into a small space at the centre of the map, while the column vertex points (with coordinates γ_{jk}) are very far out.

In the biplot it is the direction of each vertex point which is of interest, since this direction defines the line onto which the row profiles are projected.

The standard CA biplot

Exhibit 13.3:
*Standard CA biplot
of the scientific
researchers data of
Exhibit 10.1,
showing the rows in
principal
coordinates and the
columns in their
vertex directions but
rescaled by
multiplying the
standard
coordinates by the
square root of the
mass of each
column. The
position of A, for
example, in Exhibit
10.3 has been
multiplied by
$\sqrt{0.0389} = 0.197$ to
obtain the position
of A in this map.*

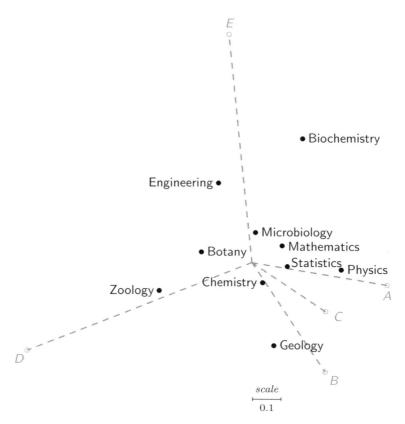

Different variations of this biplot have been proposed to redefine the lengths of these vectors. The most convenient of these alternatives is to rewrite (13.6) as follows:

$$(\frac{p_{ij}}{r_i} - c_j)/c_j^{1/2} = \sum_{k=1}^{K^*} f_{ik}(c_j^{1/2}\gamma_{jk}) + e_{ij} \tag{13.7}$$

(notice that the residuals e_{ij} in (13.7) have a different definition and standardization compared to (13.6), although we use the same notation in each case). Thus we have expressed the left-hand side as a standardized deviation from the average, and then we absorb the remaining factor $c_j^{1/2}$ into the coordinate of the column point on the right-hand side. In this way, the vertex point gets pulled inwards by an amount equal to the square root of the mass of the corresponding category, so that the rarer categories are pulled in more, which is just what we want to improve the legibility of the asymmetric map. Because this biplot represents standardized values, it is called the *standard CA biplot*. Exhibit 13.3 shows the standard CA biplot for the research funding example; compare this map with Exhibits 10.2 and 10.3. In all these maps the positions of the row points are the same, it is the positions of the column points that change (compare the scales of each map).

In Exhibit 13.3 the column points have no distance intepretation, they point in the directions of the biplot axes, and it is the projections of the row points onto the biplot axes which give estimates of the standardized values on the left-hand side of (13.7). Thus we can take a fixed reference direction, such as *D*, and then line up the projections of all the rows on this axis to estimate that Zoology has the highest profile element, then Botany, Geology, and so on, with Physics and Biochemistry having the lowest profile values on *D* (a few calculations on Exhibit 10.1 show this to be correct, with some small exceptions since this is an approximate biplot, representing 84% of the total inertia of the table).

Interpretation of the biplot

Since the projections of the rows onto the biplot axes are proportional to the values on the left-hand side of (13.7), each biplot axis can be calibrated in profile units. For example, to estimate the standardized profile values for category *A*, the projections of the row points have to be multiplied by the length of the *A* vector, equal to 0.484. To unstandardize and return to the original profile scale, multiply this length by the square root of the mass ($\sqrt{0.0389} = 0.197$) to obtain the scale factor of 0.0955. The calibration of a biplot axis is computed by simply inverting this value: $1/0.0955 = 10.47$. This is the length of the full range of one unit on the profile scale. An interval of 1% (i.e., 0.01) on the biplot axis in Exhibit 13.4 is thus a hundredth of this length, i.e., 0.1047. So we know all three facts necessary for calibrating the *A* axis: (i) the origin of the map represents the average of 0.039 (or 3.9%) for *A*); (ii) a length of 0.01 (1%) is equal to 0.1047; and (iii) the vector in Exhibit 13.3 indicates the positive direction of the axis. Exhibit 13.4 shows the calibrations on the *A* axis as well as the result of a similar calibration on the *D* axis.

Calibration of biplots

Previously we thought of the overall quality of a two-dimensional correspondence map as the amount of inertia accounted for by the first two principal axes. The biplot provides another way of thinking about the map's quality, namely as the success of recovering the profile values in the map. The original row profiles in the second table of Exhibit 10.2 can be approximately recovered by the two-dimensional biplot in Exhibit 13.3, for example by projecting all the row points onto the calibrated column axes. The closer the estimated profile values are to the true ones, the higher the quality of the map. Conversely, the differences between the true profile elements and the estimated ones can be accumulated to give an overall measure of error. When accumulated in a chi-squared fashion, i.e., by taking squared differences divided by the expected values, exactly the same measure of error will be obtained as before. In this particular example, the percentage of explained inertia in the two-dimensional map is 84%; hence the error is 16%.

Overall quality of display

Exhibit 13.4:
Symmetric map of Table 10.1 (scientific funding data), with calibrated axes for categories A and D. Notice that the calibrated axes are in the directions of the vertex points and do not pass exactly through the category profile points (in this example they come very close to the category points in principal coordinates because the difference between the inertias on the two axes is small).

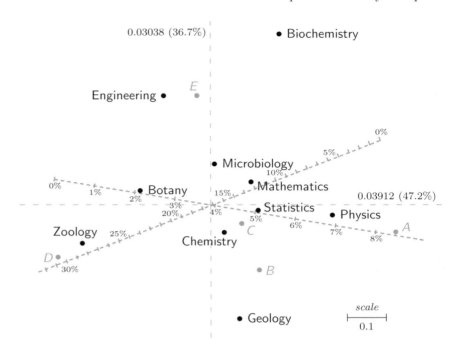

SUMMARY:
Correspondence
Analysis Biplots

1. A *scalar product* between two vectors is the product of their lengths multiplied by the cosine of the angle between them.

2. Since the perpendicular projection of a vector **x** onto the direction defined by a second vector **y** has a length equal to the length of **x** multiplied by the cosine of the angle between **x** and **y**, the scalar product can be thought of as the product of the length of the projection of **x** and the length of **y**.

3. A *biplot* is a method of displaying a point for each row and column of a data matrix in a joint map such that the scalar products between the row vectors and the column vectors approximate the values in the corresponding cells of the matrix as closely as possible.

4. Asymmetric maps in CA are biplots; strictly speaking, symmetric maps are not, although in practice the directions defined by the profile point and the corresponding vertex point are often not too different, in which case the biplot interpretation is still valid.

5. A variation of the asymmetric map which is a convenient biplot is the one where the position of each vertex point is pulled in towards the origin by an amount equal to the square root of the mass associated with the vertex category — this is the *standard CA biplot*.

6. Biplot axes passing through the origin of the map and the vertices (or rescaled vertices) may be calibrated in profile units (either proportions or percentages). This allows approximate profile values to be read directly off the map by projecting profile points onto the calibrated biplot axes.

Transition and Regression Relationships

CA produces a map where the rows and columns are depicted together as points, with an interpretation that depends on the choice amongst the many scaling options for the row and column points. Geometrically, we have seen how the positions of the row points depend on the positions of the column points, and vice versa. In this chapter we focus on the mathematical relationships between the row and column points, known as the transition equations. In addition, since regression analysis is a well-known method in statistics, we show how the row and column results and the original data can be connected through linear regression models. This chapter can be skipped without losing the thread of the presentation of the geometric interpretation of CA.

Contents

In this chapter we are interested in the relationships between all the coordinates, row and column, principal and standard, that emanate from a CA, as well as their relationships with the original data. Initially we look at the relationships that are valid for each principal axis separately. Using the scientific funding example again, we reproduce in Exhibit 14.1 all the results for the first principal axis. This axis has inertia $\lambda_1 = 0.03912$, with $\sqrt{\lambda_1} = 0.1978$. We have seen in Chapter 8 that this latter value, which is the scaling factor which links the principal to the standard coordinates, can also be interpreted as a correlation coefficient between rows and columns in terms of their coordinates on the first dimension. Since correlation is related to regression, we first look at the regression of row coordinates on column coordinates and vice versa.

Coordinates on first axis of scientific funding example

SCIENTIFIC DISCIPLINE	*Princ. coord.*	*Stand. coord.*	FUNDING CATEGORY	*Princ. coord.*	*Stand. coord.*
Geology	0.076	0.386	A	0.478	2.417
Biochemistry	0.180	0.910	B	0.127	0.643
Chemistry	0.038	0.190	C	0.083	0.417
Zoology	−0.327	−1.655	D	−0.390	−1.974
Physics	0.316	1.595	E	−0.032	−0.161
Engineering	−0.117	−0.594			
Microbiology	0.013	0.065			
Botany	−0.179	−0.904			
Statistics	0.125	0.630			
Mathematics	0.107	0.540			

*Regression
between
coordinates*

In Exhibit 8.5, using the health survey data, we showed the scatterplot of values of the row and column coordinates on the first principal axis, for each individual constituting the contingency table. Exhibit 14.2 shows the same type of plot for the standard coordinates of the scientific funding data. There are 50 points in this plot, corresponding to the 50 cells in the contingency table of Exhibit 10.1. Each box is centred on the pair of values for the respective cell, and has area proportional to the number of individuals (scientists) at that point. We know that the correlation, calculated for all 796 individuals which occur at the 50 points in this plot, is equal to 0.1978. Here we are interested in the regressions of scientific discipline on funding category, and funding category on scientific discipline. To compute a regression analysis we can string out all 796 scientists and assign their corresponding pair of values, for example a geologist in the A category would have the pair of standard coordinate values 0.386 (the y-variable, say) and 2.417 (the x-variable), according to Exhibit 14.1. Since there are only 50 unique pairs, an alternative is list the 50 pairs of coordinate values along with their frequencies and then perform a weighted regression with the frequencies as weights (this is illustrated in the Computational Appendix, pages 227–228). A standard result in simple linear regression is that the slope coefficient is equal to the correlation multiplied by the ratio of the standard deviations of the y-variable to the x-variable. The variances of the row and column standard coordinates are the same ($= 1$); hence the slope of the regression of y on x will be the same as the correlation, i.e. 0.1978 (Exhibit 14.2). In a symmetric way, the regression of x onto y will also have a slope of 0.1978, but this is with respect to the x-axis as vertical and the y-axis as horizontal — this is a slope of $1/0.1978 = 5.056$ in the plot of Exhibit 14.2 where y is the vertical axis.

*The
profile–vertex
relationship*

We saw as early as Chapter 3 that the row profiles are at weighted averages of the column vertices, where the weights are the values of the row profiles. The same relationship holds between column profiles and row vertices. These

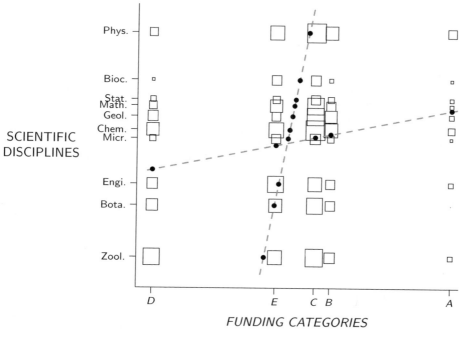

Exhibit 14.2:
Scatterplot based on standard coordinates of rows and columns on first CA dimension (Exhibit 14.1). Squares are shown at each combination of values, with area proportional to the number of respondents. The two regression lines, rows on columns and columns on rows, have slopes of 0.1978 and 5.056, inverses of each other. The dots • indicate conditional means (weighted averages), i.e., principal coordinates.

weighted average relationships hold in any projected space, in particular they hold for the coordinates along the principal axes, as illustrated in Chapter 8. In other words, along an axis k, the principal coordinates of the row points are at weighted average positions of the standard coordinates of the column points, and vice versa. This relationship can be illustrated for the first principal axis by calculating weighted average positions for each row according to the standard coordinates of the columns and vice versa, shown by the dots on the two regression lines in Exhibit 14.2. This shows that the principal coordinates lie on the two regression lines.

Regression is a model for the conditional means of the response variable with respect to the predictor variable. The dots in Exhibit 14.2 are simply the discrete sets of conditional means of y on x (five means on the line with slope 0.1978) and x on y (ten means on the line with slope 5.056). These means are the principal coordinates, which thus define the two regression functions. For example, the row of squares for **Physics** depicts the corresponding row of frequencies in the data matrix of Exhibit 10.1, plotted horizontally according to the first standard coordinates of the five column categories, that is the vertex positions on principal axis 1. The conditional mean is just the weighted average, shown by the black dot at the top of the diagram, which is thus the first principal coordinate of **Physics**. Similarly, the column of squares for

Principal coordinates are conditional means in regression

category *A* depicts the frequencies for the first column of the data matrix, plotted vertically at the first standard coordinate (vertex) positions of the ten rows. The conditional mean, shown by the black dot at the right, is then the weighted average position, which is the principal coordinate of *A*. Thus Exhibit 14.2 shows both the principal and standard coordinates together; for example, to read the principal coordinates for the rows, read the values of the ten dots on the regression line on the horizontal axis according to the scale of the standard column coordinates, and vice versa.

Simultaneous linear regressions

The fact that the regressions of *y* on *x* (rows on columns) and *x* on *y* (columns on rows) turn out to be straight lines in the CA solution is one of the oldest definitions of CA, called *simultaneous linear regressions*. If the row–column correlation is high, then the two regression lines will be closer together and the principal coordinates will have more spread; i.e., the inertia will be higher (remember that the principal inertia is the square of the correlation). In other words, CA could be defined as trying to achieve simultaneous linear regressions (i.e., a scatterplot such as Exhibit 14.2) with the least possible angle between the two regression lines, which is equivalent to maximizing the row–column correlation.

Transition equations between rows and columns

Using notation defined on pages 31 and 101, we can write these weighted average (or conditional mean) relationships between rows and columns as follows, remembering that principal coordinates correspond to profiles and standard coordinates to vertices:

$$\text{row profile} \leftarrow \text{column vertices}: \quad f_{ik} = \sum_j \left(\frac{p_{ij}}{r_i}\right) \gamma_{jk} \quad (14.1)$$

$$\text{column profile} \leftarrow \text{row vertices}: \quad g_{jk} = \sum_i \left(\frac{p_{ij}}{c_j}\right) \phi_{ik} \quad (14.2)$$

(the $\leftarrow$ stands for "is obtained from", for example "row profile $\leftarrow$ column vertices" means that the principal coordinates of a row are obtained from the standard coordinates of all the columns using the given relationship). Here we use the notation *f* and *g* for the principal row and column coordinates, γ and ϕ for the standard row and column coordinates, index *i* for rows, *j* for columns and *k* for dimensions. In parentheses we have the weights which are the row profiles in (14.1) and the column profiles in (14.2). The weighted average relationships in (14.1) and (14.2) are called the *transition equations*. Recall the relationships between principal and standard coordinates:

$$\text{row profile} \leftarrow \text{row vertex}: \quad f_{ik} = \sqrt{\lambda_k}\phi_{ik} \quad (14.3)$$

$$\text{column profile} \leftarrow \text{column vertex}: \quad g_{jk} = \sqrt{\lambda_k}\gamma_{jk} \quad (14.4)$$

where λ_k is the principal inertia (eigenvalue) on the *k*-th axis. So we could write the transition equation between row and column principal coordinates as:

$$\text{row profile} \leftarrow \text{column profiles}: \quad f_{ik} = \frac{1}{\sqrt{\lambda_k}} \sum_j \left(\frac{p_{ij}}{r_i} \right) g_{jk} \quad (14.5)$$

$$\text{column profile} \leftarrow \text{row profiles}: \quad g_{jk} = \frac{1}{\sqrt{\lambda_k}} \sum_i \left(\frac{p_{ij}}{c_j} \right) f_{ik} \quad (14.6)$$

and similarly between row and column standard coordinates:

$$\text{row vertex} \leftarrow \text{column vertices}: \quad \phi_{ik} = \frac{1}{\sqrt{\lambda_k}} \sum_j \left(\frac{p_{ij}}{r_i} \right) \gamma_{jk} \quad (14.7)$$

$$\text{column vertex} \leftarrow \text{row vertices}: \quad \gamma_{jk} = \frac{1}{\sqrt{\lambda_k}} \sum_i \left(\frac{p_{ij}}{c_j} \right) \phi_{ik} \quad (14.8)$$

Any of the above transition equations can be used trivially in a standard linear regression analysis, with the profiles as predictors, in order to "estimate" a set of coordinates. As an illustration, we recover the column standard coordinates in (14.1), using the 10×5 matrix of row profiles of Exhibit 10.1 as five predictors and the first principal coordinates of the rows (first column of Exhibit 14.1) as response. The regression analysis gives the following results for the regression coefficient:

Regression between coordinates using transition equations

Source	Coefficient
Intercept	0.000
A	2.417
B	0.643
C	0.417
D	−1.974
E	−0.161

$$R^2 = 1.000$$

The variance explained is 100% and the regression coefficients are the column standard coordinates on the first axis (see last column of Exhibit 14.1).

The more interesting and relevant regression analysis is when the data are to be predicted from the coordinates, as summarized in the CA model given in Chapter 13. We repeat this model here in three different versions, a "symmetric" version using only standard coordinates (see (13.4)), and the two asymmetric versions using row and column principal coordinates, respectively:

Recall the CA bilinear model

$$\frac{p_{ij}}{r_i c_j} = 1 + \sum_{k=1}^{K^*} \sqrt{\lambda_k} \phi_{ik} \gamma_{jk} + e_{ij} \quad (14.9)$$

$$\left(\frac{p_{ij}}{r_i} \right) / c_j = 1 + \sum_{k=1}^{K^*} f_{ik} \gamma_{jk} + e_{ij} \quad (14.10)$$

$$\left(\frac{p_{ij}}{c_j}\right)/r_i = 1 + \sum_{k=1}^{K^*} \phi_{ik} g_{jk} + e_{ij} \tag{14.11}$$

This model is called *bilinear* because it is linear in the products of parameters. We shall, however, fix either the row or column standard coordinates, and show how to obtain the principal coordinates of the other set by multiple regression analysis.

Weighted   On the left-hand sides of (14.9), (14.10) and (14.11) are the contingency ratios
regression defined in Chapter 13, written in three equivalent ways. Taking (14.10) as an example, and assuming that the standard coordinates γ_{jk} of the columns are known, we have on the right-hand side a regular regression model which is predicting the values of the rows on the left. Suppose that we are interested in the first row (Geology) and want to perform a regression for $K^* = 2$. To fit the CA model, we have to minimize a weighted sum of squared residuals, where the categories (columns) are weighted by their masses. Another way of understanding this is that in (14.10) the "predictors" γ_{jk} are normalized with respect to column masses as follows: $\sum_j c_j \gamma_{jk}^2 = 1$. Furthermore, the predictors are orthogonal as well when we weight by the column masses: $\sum_j c_j \gamma_{jk} \gamma_{j'k} = 0$ if $j \neq j'$. Hence, to perform the regression we set up the response vector as the 5×1 vector of contingency ratios for Geology, and the predictors as the 5×2 matrix of column standard coordinates on the first two principal axes. The weighted regression analysis is performed with regression weights equal to the column masses c_j. The data (contingency ratios $p_{1j}/(r_1 c_j)$ for Geology (row 1), γ_1 and γ_2 denoting column standard coordinates for dimensions 1 and 2 respectively, and weights c_j) are as follows:

Category	Geology	γ_1	γ_2	Weight
A	0.9063	2.4175	-0.4147	0.0389
B	1.3901	0.6434	-0.9948	0.1608
C	1.1781	0.4171	-0.2858	0.3894
D	1.0163	-1.9741	-0.7991	0.1621
E	0.4730	-0.1613	1.6762	0.2487

The results of the regression are:

Source	Coefficient	Standardized coefficient
Intercept	1.000	—
f_{11}	0.076	0.234
f_{12}	-0.303	-0.928

$$R^2 = 0.916$$

The coefficients are the principal coordinates f_{11} and f_{12} of Geology (see the first one in Exhibit 14.1) while the variance explained (R^2) is the quality of Geology in the two-dimensional map (see Exhibit 11.8).

Since the predictors are standardized and orthogonal in the weighted regression, it is known that the standardized regression coefficients are also the partial correlations between the response and the predictors. The correlation matrix between all three variables is as follows (remember that the weights are included in the calculation):

Variables	Geology	γ_1	γ_2
Geology	1.000	0.234	-0.928
γ_1	0.234	1.000	0.000
γ_2	-0.928	0.000	1.000

The two predictors are uncorrelated, as expected, and the correlations between Geology and the two predictors are exactly the standardized regression coefficients. The squares of these correlations, $0.234^2 = 0.055$ and $(-0.928)^2 = 0.861$, are the squared cosines (relative contributions) given in Exhibit 11.6. The above series of results illustrates the property in regression that if the predictors are uncorrelated, then the variance explained R^2 is equal to the sum of squares of the partial correlations.

The transition equations (14.1) and (14.2) are the basis of a popular algorithm for finding the solution of a CA, called *reciprocal averaging*. The algorithm starts from any set of standardized values for the columns, say, where centring and normalizing are always with respect to weighted averages and weighted sum of squares. Then the averaging in (14.1) is applied to obtain a set of row values. The row values are then used in the averaging equation of (14.2) to obtain a new set of column values. The column values are restandardized (otherwise the sucessive averagings would just collapse the values to zero). The above process is repeated until convergence, giving the coordinates on the first principal axis. Finding the second set of coordinates is more complicated because we have to ensure orthogonality with the first, but the idea is the same. We have shown in different ways that the passage from column to row coordinates and row to column coordinates can be described by a regression analysis in each case, so that this flip-flop process is also known as *alternating least-squares*, or alternating regressions. Numerically, it is better to perform the computations using the SVD (see page 47 as well as the Theoretical and Computational Appendices), but for a fuller understanding of CA it is illuminating to be aware of these alternative algorithms.

SUMMARY:
Transition and
Regression
Relationships

1. For any values assigned to the row and column categories, the conditional means (i.e., regressions) can be computed for rows on columns or columns on rows.

2. The CA solution, using standard coordinates on a particular axis as the two sets of values, has the following properties:
   - the two regressions are linear (hence the name *simultaneous linear regressions*);
   - the angle between the two regression lines is minimized;
   - the conditional means which lie on the two regression lines are the principal coordinates.

3. The weighted average relationship between row and column coordinates, when the weights are the elements of profiles (row or column profiles as the case may be) are called *transition equations*. Successive applications of the pair of transition equations leads to an algorithm for finding the CA solution, called *reciprocal averaging*.

4. CA can be defined as a *bilinear regression model*, since the data can be recovered by a model that is linear in products of the row and column coordinates. This model becomes linear if either set of coordinates is regarded as fixed, leading to an algorithm for finding the CA solution called *alternating least-squares regressions* (which, in fact, is identical to the reciprocal averaging algorithm).

Clustering Rows and Columns

Up to now we have been transforming data matrices to maps where the rows and columns are displayed as points in a continuous space, usually a two-dimensional plane. An alternative way of displaying structure consists in performing separate cluster analyses on the row and column profiles. This approach has close connections to CA and decomposes the inertia according to the discrete groupings of the profiles rather than along continuous axes. In the case of a contingency table there is an interesting spin-off of this analysis in the form of a statistical test for significant clustering of the rows or columns.

Contents

Partitioning the rows or the columns

The idea of grouping objects together is omnipresent in data analysis. The grouping might be a given classification, or it might be determined according to some criterion which clusters similar objects together. We first consider the former case, when the grouping is established according to a categorical variable which classifies the rows or columns of a table. Taking the scientific research funding example again, suppose that there is a pre-determined grouping of the scientific disciplines into four groups, according to university faculties: {Geology, Physics, Statistics, Mathematics}, {Biochemistry, Chemistry}, {Zoology, Microbiology, Botany} and {Engineering}. As we pointed out in Chapter 12, when a categorical variable is defined on the rows, as in this example, each category defines a supplementary row of the table which merges the frequencies of the rows indicated by that category. Thus the ten rows of Exhibit 10.1 are condensed into four rows corresponding to the four groups, shown in Exhibit 15.1. The CA of the original data of Exhibit 10.1 had a total inertia

113

Exhibit 15.1:
*Frequencies of
funding categories
for 796 researchers
grouped into four
categories according
to scientific
discipline.*

| SCIENTIFIC GROUPS | FUNDING CATEGORIES | | | | | |
	A	B	C	D	E	*Sum*
Geol/Phys/Stat/Math	17	57	134	35	63	*306*
Bioc/Chem	7	27	62	22	41	*159*
Zool/Micr/Biol	4	33	89	57	60	*243*
Engi	3	11	25	15	34	*88*
Sum	*31*	*128*	*310*	*129*	*198*	*796*

of 0.08288, whereas if we perform the CA of Exhibit 15.1, the total inertia turns out to be 0.04386. There is a loss of inertia when points are merged, or putting this the other way around, there is an increase in inertia if a row or column is split apart according to some subclassification.

Between- and within-groups inertia

The inertia of the merged table in Exhibit 15.1 is called the *between-groups inertia*, since it measures the variation in the table between the four groups of rows. The difference between the total inertia of 0.08288 and the between-groups inertia of 0.04386 is called the *within-groups inertia*, measuring the variation within the four groups which is lost when we merge the rows into the groups. This decomposition of inertia is a classic result of analysis of variance, usually applied to a single variable, but equally applicable to multivariate data. In the CA context each row profile, denoted by $\mathbf{a}_i$, has a mass r_i assigned to it, and the average row profile (centroid) is the vector $\mathbf{c}$ of column masses. Distances between row profiles are measured by the χ^2-distance, for example let d_i denote the χ^2-distance between $\mathbf{a}_i$ and $\mathbf{c}$. Then the total inertia is $\sum_i r_i d_i^2$ (formula (4.7), page 29). Between-group inertia is a similar formula, but applied to the merged rows as follows. Suppose $\bar{\mathbf{a}}_g$ denotes the profiles of the merged rows, where $g = 1, \ldots, G$ is the index of the groups (here $G = 4$), and the mass of the g-th group, $\bar{r}_g$, is the sum of the masses of the members of the group. The profiles $\bar{\mathbf{a}}_g$ still have centroid at $\mathbf{c}$ and, denoting their χ^2-distances to the centroid by $\bar{d}_g$, the between-group inertia is $\sum_g \bar{r}_g \bar{d}_g^2$. Finally, each group g has an inertia with respect to its own centroid $\bar{\mathbf{a}}_g$: if d_{ig} denotes the χ^2-distance from each profile i in group g to the centroid $\bar{\mathbf{a}}_g$, then the inertia within the g-th group is $\sum_{i \epsilon g} r_i d_{ig}^2$, where $i \epsilon g$ means the set of rows in group g. Summing this quantity over all the groups gives the within-groups inertia. The decomposition of inertia is thus:

total inertia = between-groups inertia + within-groups inertia

$$\sum_i r_i d_i^2 = \sum_g \bar{r}_g \bar{d}_g^2 + \sum_g \sum_{i \epsilon g} r_i d_{ig}^2 \qquad (15.1)$$

$$0.08288 = 0.04386 + 0.03902$$

The within-groups inertia is equal to 0.03902, according to the above, but what is the contribution from each of the four groups? One can calculate this directly, remembering to use the same values of **c** in the χ^2-distance in all the calculations, but a quicker way is to apply CA to a matrix where, one at a time, we merge the groups. For example, if we merge Geology, Physics, Statistics and Mathematics into the first group and then analyse this merged row along with the other (unmerged) rows (i.e., seven rows in total), the total inertia is 0.06446. Compared to the total inertia 0.08288 of the original data set, the reduction by 0.01842 represents the within-group inertia for that group which was lost in the merging. Then we merge Biochemistry and Chemistry and analys e the matrix with six rows, and the inertia now drops to 0.05382, so the within-groups inertia of that group is the difference, $0.06446 - 0.05382 = 0.01064$ and so on. Exhibit 15.2 gives the complete decomposition of inertia, in raw units and percentages. Notice that the within-groups inertia of the group composed of one row, Engineering, is 0.

Group	Definition	Component	% of part	% of total
Between-groups inertia:				
Geol/Phys/Stat/Math	$\bar{r}_1 \bar{d}_1^2$	0.01482	33.8%	17.9%
Bioc/Chem	$\bar{r}_2 \bar{d}_2^2$	0.00099	2.3%	1.2%
Zool/Micr/Bota	$\bar{r}_3 \bar{d}_3^2$	0.01548	35.3%	18.7%
Engi	$\bar{r}_4 \bar{d}_4^2$	0.01256	28.6%	15.2%
Total	$\sum_g \bar{r}_g \bar{d}_g^2$	0.04386	100.0%	52.9%
Within-groups inertia:				
Geol/Phys/Stat/Math	$\sum_{i \in 1} r_i d_{i1}^2$	0.01842	47.2%	22.2%
Bioc/Chem	$\sum_{i \in 2} r_i d_{i2}^2$	0.01064	27.3%	12.8%
Zool/Micr/Bota	$\sum_{i \in 3} r_i d_{i3}^2$	0.00996	25.5%	12.0%
Engi	$\sum_{i \in 4} r_i d_{i4}^2$	0	0%	0%
Total	$\sum_g \sum_{i \in g} r_i d_{ig}^2$	0.03902	100.0%	47.1%

Exhibit 15.2: *Decomposition of inertia between and within groups, including components of each part and percentages with respect to the part and the total. The sum of the total of the between-groups inertia and the total of the within-groups inertia is the total inertia 0.08288 of the original table (Exhibit 10.1).*

In the above, the partition of the rows into groups was given by available information, but we now consider constructing groups using a particular type of cluster analysis. We use a small data matrix to illustrate the calculations involved. This example is taken from an actual sample of 700 shoppers at five different food stores. The sample has been tabulated according to store and age group, yielding the 5×4 table in Exhibit 15.3. The χ^2 statistic for this table is 25.06, which corresponds to a P value of 0.015. Thus we would conclude that there exists a significant association between age and choice of store. Alongside the table we show the symmetric CA map. A market

Data set 8: Age distribution in food stores

Exhibit 15.3:
*Cross-tabulation of
food stores by age
groups, for 700
consumers, and
symmetric CA map
which explains
97.2% of the inertia.*

| FOOD | AGE GROUP (years) | | | | |
STORE	16-24	25-34	35-49	50+	Sum
A	37	39	45	64	185
B	13	23	33	38	107
C	33	69	67	56	225
D	16	31	34	22	103
E	8	16	21	35	80
Sum	107	178	200	215	700

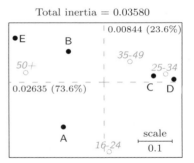

Total inertia = 0.03580

researcher would be interested to know where this significant association is concentrated; for example, which stores or group of stores have a significantly different age profile from the others. The major contrast in the data is between the oldest group on the left and the second youngest group on the right. Store E is the most associated with the oldest group and stores C and D tend more towards the younger ages. The vertical axis contrasts the youngest age group with the others. Store A appears to separate from the others towards the youngest age group.

*Clustering
algorithm*

We now construct a partition of the rows and columns using a clustering algorithm which tries to maximize the between-groups inertia and — simultaneously — minimize the within-groups inertia. The clustering algorithm is illustrated in Exhibit 15.4 for the rows. At the start of the process, each row is separate and the between-groups inertia is just the total inertia. Any merging will reduce the between-groups inertia, so the first step is to identify which pair of rows (stores) can be merged to result in the least reduction in the inertia. The two rows which are the most similar in this sense are stores C and D. When these rows are merged to form a new row, labelled (C,D), the inertia for the resultant 4×4 table is reduced by 0.00084 to 0.03496, or on the χ^2 scale by 0.59 to 25.06 (in Exhibit 15.4 we report the χ^2 values, which are always the inertia values multiplied by the sample size $N = 700$: $\chi^2 = 0.03496 \times 700 = 25.06$). In percentage terms this is a decrease of 2.3% in χ^2 or in inertia. The procedure is then repeated to find the rows in the new table which are the most similar in this sense. These turn out to be stores B and E, leading to a further reduction in χ^2 of 1.53 (6.1%). The table now has three rows labelled A, (B,E) and (C,D). The procedure is repeated and the smallest reduction is found when store A joins the pair (B,E) to form a new row labelled (A,B,E), reducing χ^2 by a further 5.95 (23.8%). Finally, the two rows (A,B,E) and (C,D) merge to form a single row, consisting of the marginal column sums of the original table, for which the χ^2 is zero. The final reduction is thus 16.99 (67.8%), which was the inertia of the penultimate table in Exhibit 15.4. The whole procedure can be repeated on the columns of the table in an identical fashion.

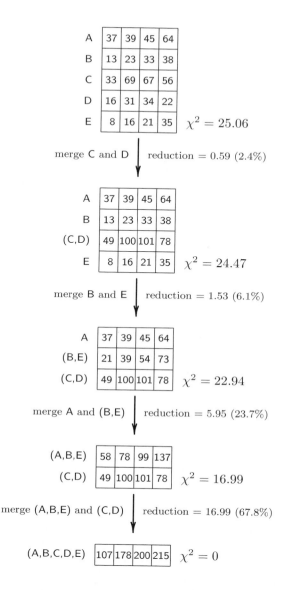

Exhibit 15.4:
Steps in the clustering of the rows of Exhibit 15.1: at each step two rows are merged, chosen to induce the minimum decrease in the χ^2-statistic, or equivalently in the between-group inertia (to convert the χ^2 values to inertias, divide by the sample size $N = 700$).

The successive merging of the rows, called *hierarchical clustering*, can be depicted graphically as a *binary tree* or *dendrogram* — this is shown in Exhibit 15.5 along with a similar hierarchical clustering of the columns. Notice that the ordering of the rows and columns of the original table usually requires modification to accommodate the tree displays, although in this particular example only the rows need to be reordered. The fact that stores C and D are the first to merge is apparent on the tree. The point at which this merging

Tree representations of the clusterings

occurs is called a *node*, and in each case the level of the node corresponds to the associated reduction in χ^2.

Decomposition of inertia (or χ^2)

Since the original χ^2 statistic is reduced to zero at the end of the clustering process, it is clear that the set of reductions forms a decomposition of χ^2: $25.06 = 16.99 + 5.95 + 1.53 + 0.59$. Dividing by the sample size 700 gives the corresponding decomposition of inertia: $0.03580 = 0.02427 + 0.00851 + 0.00218 + 0.00084$. The percentage form is the same for both decompositions: 67.8%, 23.8%, 6.1% and 2.3%. The columns are merged in an identical fashion and the values of the nodes again constitute a decomposition of inertia (or χ^2): $0.03580 = 0.02383 + 0.00938 + 0.00259$, or in percentage form 66.6%, 26.2% and 7.2%.

Deciding on the partition

In cluster analyses of this type, the trees are inspected to deduce the number of clusters of objects. For example, looking at the row clustering we see that there is a large difference between the two clusters of food stores (C,D) and (A,B,E), indicated by the high value at which these clusters merge. Thanks to the decomposition of inertia, we could say that 67.8% of the inertia is accounted for if we condense the rows into these two clusters. If we separate store A as a third cluster, then a further 23.7% of the inertia is accounted for, i.e. 91.5%. Percentages of inertia associated with the nodes of such a cluster analysis are thus interpreted in much the same way as percentages of inertia of principal axes in CA. The decision as to what percentage is great enough to halt the interpretation is usually an informal one, based on the sequence of percentages and the substantive interpretation of each node or principal axis.

Testing hypotheses on clusters of rows or columns

The χ^2 statistic for the original contingency table was reported to be significant ($P = 0.015$); hence somewhere in the table there must be significant differences amongst the profiles. To pinpoint which profiles are significantly different in a statistical sense is not a simple question because there are many possible groupings of the stores that we could test and the significance level has to be adjusted if many tests are performed on the same data set. Furthermore, particular groupings, for example of stores C with D and B with E, have been suggested by the data themselves and have not been set up as hypotheses before the data were collected.

Multiple comparisons

Here we are treading the fine line between exploratory and confirmatory data analysis, trying to draw statistical conclusions from data that were collected in an exploratory fashion with no fixed *a priori* hypotheses. Fortunately, an area of statistical methodology has been developed specially for this situation, called *multiple comparisons*. This approach is often used in the analysis of experiments where there are several "treatments" being tested in one experiment, rather than the classic single treatment versus control situation. A multiple comparisons procedure allows any treatment (or group of treat-

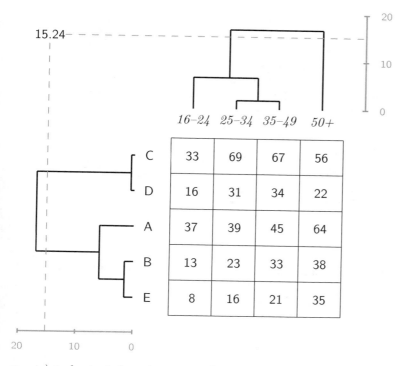

Exhibit 15.5:
The tree structures depicting the hierarchical clusterings of the rows and columns. The clustering is in terms of χ^2, and can be converted to inertias by dividing by the sample size, 700. The critical level of 15.24 on the χ^2 scale is indicated, for both rows and columns.

ments) to be tested against any other, and statistical decisions may be made at a prescribed significance level to protect all these tests from the so-called "Type I Error", i.e., finding a result which has arisen purely by chance.

As in the case of different treatments in an experimental situation, we would like to test the differences between any two rows, say, of the table or any two groups of rows. If there were only one test to do, we would calculate the reduced table consisting of the two rows (or merged groups) and make a one-off χ^2 test in the usual way. The multiple comparisons procedure developed for this situation allows testing for differences between any two rows (or groups of rows). The usual χ^2 statistic for the reduced table is calculated but compared to different critical values for significance. In the Theoretical Appendix, Exhibit A.1 on page 211, we give a table of critical points for this test, at the 5% significance level, for contingency tables of different sizes. In our 5×4 example the critical point can be read from the table as 15.24: so if the χ^2 statistic is superior to 15.24, then it can be deduced that the two rows (or groups of rows) are significantly different.

Multiple comparisons for contingency tables

The critical value for the multiple comparison test can be used for any subset or merging of the rows or columns of the table, in particular it can be used for our hierarchical clusterings, allowing us to separate out the statistically

Cut-off χ^2 value for significant clustering

significant groups, as shown in Exhibit 15.5. The interpretation of this cut-off point is that, amongst the age groups, it is really the contrast between the oldest age group and the rest that is statistically significant; and, concerning the food stores, the statistical differences lie between two groups, (A,B,E) and (C,D). Thus the contrast observed along the second axis of Exhibit 15.3 could be due to random variation in the observed data — the separation of the youngest age group from the others is not significant, and the distinction between age groups *16–24* and *35–49* years along the second axis is equally difficult to justify from a statistical point of view. This does not mean, of course, that we are prevented from inspecting the original information in the form of the two-dimensional map of Exhibit 15.3 — the data content is always worth considering irrespective of the statistically significant features. In Chapter 25 we will use these same critical values for a significance test on the principal inertias of a contingency table.

Ward clustering

The clustering algorithm described in this chapter is a special case of *Ward clustering*. In this type of clustering, clusters are merged according to a minimum-distance criterion which takes into account the weights of each point being clustered. So, instead of thinking of this as a reduction in χ^2 (or inertia) at each step, the χ^2-distances between the profiles could be used and the associated masses. The "distance" between two row clusters g and h, for example, is then

$$\frac{\bar{r}_g \bar{r}_h}{\bar{r}_g + \bar{r}_h} \|\bar{\mathbf{a}}_g - \bar{\mathbf{a}}_h\|_c^2 \tag{15.2}$$

where $\bar{r}_g$ and $\bar{r}_h$ are the masses of the respective clusters, and $\|\bar{\mathbf{a}}_g - \bar{\mathbf{a}}_h\|_c$ is the χ^2-distance between the profiles of the groups.

SUMMARY:
Clustering the
Rows and Columns

1. Cluster analyses of the rows or columns provide an alternative way of looking for structure in the data, by collecting together similar rows (or columns) in discrete groups.

2. The results of the clusterings can be depicted graphically in a tree structure (*dendrogram* or *binary tree*), where nodes indicate the successive merging of the rows (or columns).

3. The total inertia (or equivalently the χ^2 statistic) of the table is reduced by a minimum at each successive level of merging of the rows (or columns). This *Ward clustering* procedure provides a decomposition of inertia with respect to the nodes of the tree, analogous to the decomposition of inertia with respect to principal axes in correspondence analysis.

4. Thanks to a *multiple comparisons* procedure, the inertia component accounted for by each node can be tested for significance, leading to statistical statements of difference between groups of rows (or groups of columns). This test applies to valid contingency tables only.

Multiway Tables

Up to now we have dealt exclusively with two-way tables in which the frequencies of co-occurrence of two variables have been mapped. We now start to consider situations where data are available on more than two variables and how we can explore such data graphically. One approach is to re-express the multiway frequency table in the form of a two-way table and to use the usual simple CA approach.

Contents

We return to the health assessment data (data set 3) that were discussed at length in Chapters 6 and 7, namely the representative sample of 6371 Spaniards cross-tabulated by age and self-assessment of their own health (Exhibit 6.1). Several other variables are available, for example gender, education, region of residence, and so on. We use the simplest of these, gender with two categories, as an example of how to introduce a third variable into the CA. Two further cross-tabulations can now be made: gender by age group and gender by health category. While the former table might be interesting from a demographic point of view, the latter table is more relevant to the substantive issue of health assessment — see Exhibit 16.1. There is no need to perform a CA of this table to see the pattern in the numbers — this 2×5 table is inherently one-dimensional and all the results are in the percentages. It is clear that males generally have a better opinion of their health; there are higher percentages of males in the *very good* and *good* categories, while the females are higher in the *regular*, *bad* and *very bad* categories.

Introducing a third variable in the health assessment data

We saw previously in Chapter 6 that self-perceived health deteriorated with age. Separately, Exhibit 16.1 shows a gender-related effect, with men on average more optimistic about their health than women. The question now is whether the gender effect is the same across all age groups or whether it

Interaction between variables

121

Exhibit 16.1:
Cross-tabulation of gender with self-perceived health, showing row profile values as percentages. Data source: Spanish National Health Survey, 1997.

GENDER	Very Good	Good	Regular	Bad	Very Bad	Sum
male	448	1789	636	177	39	3089
%	14.5	57.9	20.6	5.7	1.3	
female	369	1753	859	237	64	3282
%	11.2	53.4	26.2	7.2	2.0	
Sum	817	3542	1495	414	103	6371
%	12.8	55.6	23.5	6.5	1.6	

changes; for example, it could be that in a particular age group the gender effect is greater or is even reversed. This phenomenon is called an *interaction*, in this case an interaction between age and gender. Absence of an interaction would mean that the same gender difference exists across all age groups.

Interactive coding

To be able to visualize possible interactions between gender and age, we need to code the data in a more detailed manner. A new variable is created of all the combinations of gender and age; in this case 2 genders and 7 age groups give $2 \times 7 = 14$ combinations in total — this process is called *interactive coding*. The interactively coded variable is then cross-tabulated with the health categories to give the contingency table of Exhibit 16.2.

CA of the interactively coded cross-tabulation

Exhibit 16.3 shows the symmetric map of Exhibit 16.2. Here the two-dimensional map is given although the result is still highly one-dimensional as we saw in Chapter 6. In this map there are two points showing male–female differences across the age groups. Comparing the pairs of gender points for each age group we see consistently that the female point is to the left of the male counterpart, illustrating the effect in Exhibit 16.1 that females are generally less optimistic about their health. There is no reversing of this phenomenon in any age group; however, there are some differences in the distances between the male and female points. At the younger ages the male–female distances are relatively small, up to the 35-44 age group. In the 45-54 age group, where large changes appear in self-perceived health (see Chapter 6), there is also a bigger difference between men and women. This change is maintained in the older age groups, and we even see that females in the 55-64 age group are more pessimistic than males in the higher 65-74 age group. Similarly, females in age group 65-74 are more pessimistic than males in the older group 75+. This changing difference between men and women across the age groups is evidence of a gender-age interaction when it comes to self-assessment of health.

Data set 9: Opinions about working women

As another illustration of interactive coding, we now introduce a data set that we shall be using several times in this and following chapters. These data are taken from the International Social Survey Programme (ISSP) survey of

GENDER–AGE	Very Good	Good	Regular	Bad	Very Bad	Sum
m16-24	145	402	84	5	3	*639*
m25-34	112	414	74	13	2	*615*
m35-44	80	331	82	24	4	*521*
m45-54	54	231	102	22	6	*415*
m55-64	30	219	119	53	12	*433*
m65-74	18	125	110	35	4	*292*
m75+	9	67	65	25	8	*174*
f 16-24	98	387	83	13	3	*584*
f 25-34	108	395	90	22	4	*619*
f 35-44	67	327	99	17	4	*514*
f 45-54	36	238	134	28	10	*446*
f 55-64	23	195	187	53	18	*476*
f 65-74	26	142	174	63	16	*421*
f 75+	11	69	92	41	9	*222*

Exhibit 16.2:
Cross-tabulation of interactively coded gender–age variable with self-perceived health (m=male, f=female, seven age groups as in Exhibit 6.1). Each row of Exhibit 6.1 has been subdivided into two rows according to gender.

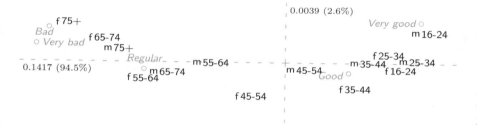

Exhibit 16.3:
Symmetric CA map of interactively coded gender–age variable cross-tabulated with health categories; the gender–age profiles are situated at the positions of the m and f labels.

Family and Changing Gender Roles in 1994, involving a total sample of 33,590 respondents and conducted in 24 countries (former East and West Germany are still considered separately in the ISSP surveys, as are Great Britain and Northern Ireland). For our purposes here we consider the relationships between some demographic variables and the responses to the following question related to women's participation in the labour market: "Considering a woman who has a schoolchild at home, should she work full-time, work part-time, or stay at home?" As in all such questionnaire surveys, there is an additional response option "unsure/don't know" to which we have also added the few non-responses (dealing with non-responses will be discussed in more detail in Chapter 21). In addition to the responses to this question, we have data on several demographic variables for each respondent, of which the following three will be of interest here: gender (2 categories), age (6 categories) and country (24 categories). The response frequencies for each country are given in Exhibit 16.4.

Exhibit 16.4:
Frequencies of
response to question
on women working
when they have a
schoolchild at home,
for 24 countries
(Source: ISSP
survey on Family
and Changing
Gender Roles, 1994;
West and East
Germany are still
kept separate), with
the average profile
in percentage form.
The following
abbreviations are
used: W=work
full-time, w=work
part-time, H=stay
at home, ?=don't
know/unsure/missing.

COUNTRIES		*W*	*w*	*H*	*?*	*Sum*
AUS	Australia	256	1156	176	191	*1779*
DW	West Germany	101	1394	581	248	*2324*
DE	East Germany	278	691	62	66	*1097*
GB	Great Britain	161	646	70	107	*984*
NIRL	Northern Ireland	126	394	75	52	*647*
USA	United States	482	686	107	172	*1447*
A	Austria	84	632	202	59	*977*
H	Hungary	285	736	447	32	*1500*
I	Italy	171	670	167	10	*1018*
IRL	Ireland	223	424	209	82	*938*
NL	Netherlands	539	1205	143	81	*1968*
N	Norway	487	1242	205	153	*2087*
S	Sweden	295	833	39	105	*1272*
CZ	Czechoslovakia	228	585	198	13	*1024*
SLO	Slovenia	341	428	222	41	*1032*
PL	Poland	431	425	589	152	*1597*
BG	Bulgaria	270	427	335	94	*1126*
RUS	Russia	175	1154	550	119	*1998*
NZ	New Zealand	120	754	72	101	*1047*
CDN	Canada	566	497	108	269	*1440*
RP	Phillipines	243	448	484	25	*1200*
IL	Israel	468	664	92	63	*1287*
J	Japan	203	671	313	120	*1307*
E	Spain	738	1012	514	230	*2494*
Sum		7271	17774	5960	2585	*33590*
%		*21.6%*	*52.9%*	*17.7%*	*7.7%*	

Basic CA map
of countries by
responses

The CA map of this table is shown in Exhibit 16.5 (here we change the graphical style of our CA maps — we shall comment on different software options for producing the maps at the end of the Computational Appendix). The interpretation of this map is quite clear; the contrast from left to right is between women working (on the left) versus women staying at home (on the right), while the vertical contrast is between women working full-time (at the top) versus women working part-time (at the bottom). Countries such as the Phillipines and Poland are the most traditional on this issue, whereas countries such as Sweden, East Germany, Israel, New Zealand, Great Britain and Canada are the most liberal. On the left, the difference between the countries in the vertical direction separates out those like Canada who are the most in favour of women working full-time versus New Zealand, for example, more in favour of part-time employment. Remember that the origin of the map represents the average profile in the last row of Exhibit 16.4, so that all countries on the left are more liberal than average, while if two countries

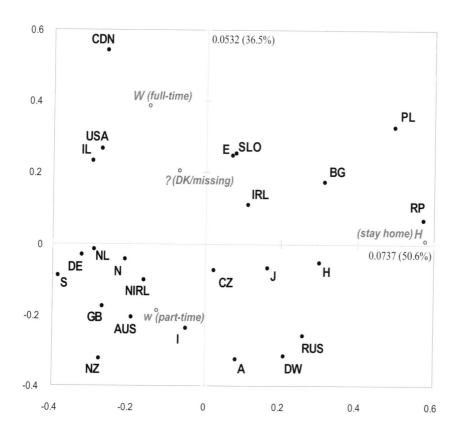

Exhibit 16.5:
*Symmetric CA map
of 24 countries and
4 question response
categories (Exhibit
16.4).*

are at the same position on the horizontal axis (for example, USA and Great
Britain) the country more positive on the vertical axis will be more in favour
of women working full-time than part-time.

We first interactively code gender with country in order to visualize male–
female differences. Exhibit 16.6 shows the first and last rows of the 48 × 4
contingency table. The map in Exhibit 16.7 has not changed much in terms
of the positions of the reponse categories, but it is interesting to compare the
pairs of points for each country. In almost all cases the female point is more
to the left compared to the male counterpart (Bulgaria is the only exception).
Attitudes within a country are surprisingly homogeneous compared to the
large between-country differences. The countries where there is the biggest
distance between male and female opinion are mostly on the conservative
side of the map, for example the Phillipines, Japan, Northern Ireland, West
Germany and Spain, while on the left side of the map Australia shows one
of the biggest male–female differences. In this analysis the inertia must be
higher than the previous one since the splitting of the samples by gender

*Introducing gender
interactively*

Exhibit 16.6:
Frequencies of response to question on women working when they have a pre-school child at home, for 24 countries, i.e., Exhibit 16.4 split by gender (any slight discrepancies between the subtotals for a country and the totals in Exhibit 16.4 are due to a few missing values for gender).

COUNTRY	*W*	*w*	*H*	*?*	*Sum*
AUSm	117	596	114	82	*909*
AUSf	138	559	60	109	*866*
DWm	43	675	357	123	*1198*
DWf	58	719	224	125	*1126*
DEm	146	316	29	37	*528*
DEf	132	375	33	29	*569*
...	...	...	...	...	...
...	...	...	...	...	...
Ilm	220	275	57	29	*581*
Ilf	247	387	35	34	*703*
Jm	85	279	171	57	*592*
Jf	118	392	142	63	*715*
Em	347	445	294	111	*1197*
Ef	390	566	218	118	*1292*

must add inertia; in fact, the total inertia in the present analysis is 0.01546, whereas in the previous one it was 0.01456. Thus we can say that the part due to gender difference in the present analysis is 0.00090, which is 5.8% of the inertia. As can be seen in the map, the major differences are between countries, not between genders.

Introducing age group and gender

Since the sample sizes in each country are so large, we can split the samples even further by age, that is each country-gender group is subdivided into six age groups: up to 25 years old, 26-35, 36-45, 46-55, 56-65, and 66+ years. Hence we code interactively three variables into one, with $24 \times 2 \times 6 = 288$ categories in total. The CA of the resultant 288×4 table is shown in Exhibit 16.8, and again remains remarkably stable as far as the response categories are concerned. The 288 row points are represented by dots since it is impossible to label each one. Some outlying points are labelled; for example, the most liberal group lying far out at top left is the youngest group of female Canadians up to 25 years old. Of this subsample of 168 women, 101 (60.1%) are in favour of women with a schoolchild at home working full-time, 32 (19.0%) respond part-time, 3 (1.8%) say women should stay at home, and 32 (19.0%) do not respond or are missing (as we shall see in Chapter 21, there are a lot of "don't knows" in the Canadian sample as a whole). The most liberal male group is the youngest East German male group. At the other extreme on the right we have the oldest group of Hungarian and Polish males; for example, of the 76 Polish men 66 years or older, 16 (21.1%) respond full-time, 13 (17.1%) part-time, 41 (53.9%) stay at home, with a non-response of 6 (7.9%). At the bottom we have the oldest group of New Zealand males — these will be the most in favour of part-time work.

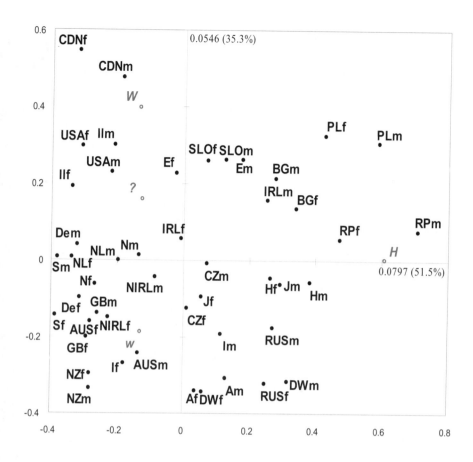

Exhibit 16.7:
Symmetric CA map of interactively coded data (Exhibit 16.6). The male points are consistently to the right of their female counterparts, with the exception of Bulgaria (BG), where females have a more conservative attitude than males.

Finally, notice the curve of the cloud of points in Exhibit 16.8, called the *arch effect* or the *horseshoe*. This commonly found phenomenon is a result of the profile space being a simplex, in the present case a tetrahedron in three dimensions since there are four columns. Any gradient of change from one extreme corner of the space (*W* — work fulltime) to another (*H* — stay at home) will follow a curved path in this restricted space, rather than a straight line. Points that lie inside the arch, such as the labelled group of Polish males 26–35 years old, will tend to be polarized in the sense of being high on the two extreme responses. Of the 141 respondents in this group, 45 (31.9%) respond full-time, 31 (22.0%) part-time, 45 (31.9%) stay at home, and 20 did not respond (14.2%) — so this group does have above average responses on both extremes.

Arch ("horseshoe") pattern in the map

Exhibit 16.8:
*Symmetric CA map
of three-way
interactively coded
data. The
country–gender–age
groups are
represented by dots,
and form a curved
pattern that is
encountered
frequently in CA
maps when the
profiles fall on a
gradient from one
extreme (W) to the
other (H).*

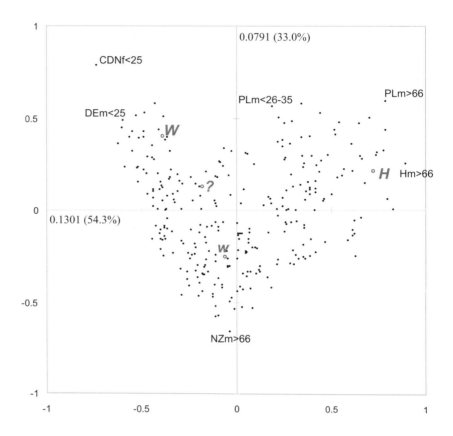

SUMMARY:
Multiway Tables

1. Two or more categorical variables can be *interactively coded* into a new variable which consists of all combinations of the categories. For example, two variables with J_1 and J_2 categories would be coded into a new variable with $J_1 J_2$ categories.

2. CA is applied to an interactively coded variable that is cross-tabulated with another variable. The resulting map shows the interaction pattern between the variables that have been interactively coded.

3. Interactive coding of multiway tables would normally not proceed beyond three variables interactively coded, since the number of categories increases rapidly, as well as the complexity of the map. The level of interaction that can be investigated depends on how much data are available, because interactive coding fragments the sample into small subsamples, and these subsamples should not be too small.

Stacked Tables

Survey research in the social sciences usually involves a multitude of variables. For example, in a questionnaire survey there are many question responses as well as many demographic characteristics which we want to relate to respondents' attitudes. The advantage of CA is the ability to visualize many variables simultaneously, but there is a limit to the number of variables that can be interactively coded, as illustrated in the previous chapter, owing to the large number of category combinations. When there are many variables an alternative procedure is to code the data in the form of *stacked*, or *concatenated*, tables. The relationship between each demographic variable and each attitudinal variable can then be interpreted in a joint map. In this chapter we give examples of this approach, both when there are several demographic characteristics and when there are responses to several questions.

Contents

We now expand the data set from Chapter 16 on attitudes toward women working by including, in addition to country (24 categories, see Exhibit 16.4 for abbreviations), gender (2 categories, M and F) and age group (6 categories, A1 to A6), the two variables marital status (5 categories) and education level (7 categories), totalling five demographic variables. The definitions and abbreviations of the two additional variables are as follows:

Several demographic variables, one question

— *Marital status:* ma (married), wi (widowed), di (divorced), se (separated), si (single)

— *Education:* E1 (no formal education), E2 (incomplete primary), E3 (primary), E4 (incomplete secondary), E5 (secondary), E6 (incomplete tertiary), E7 (tertiary)

Exhibit 17.1:
*Stacking of
contingency tables
which separately
cross-tabulate five
demographic
variables with the
responses to the
question on women
working
(W=full-time,
w=part-time,
H=stay at home,
?=don't know/non-
response).*

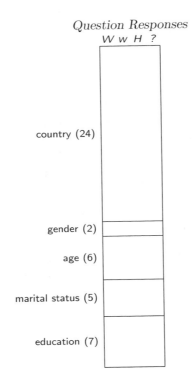

*Stacking as an
alternative to
interactive coding*

It is clearly not possible to code interactively all five variables: the number of combinations would be $24 \times 2 \times 6 \times 5 \times 7 = 10,080$ combinations! As an alternative, we can cross-tabulate each demographic variable with the responses and stack the contingency tables on top of one another, as depicted in Exhibit 17.1. The top table is the one in Exhibit 16.1, with countries as rows, then the table with two rows for gender, then six rows for age group and so on, constituting a table with $24 + 2 + 6 + 5 + 7 = 44$ rows, one for each demographic category. This type of coding will not reveal interactions and should be regarded as a type of average CA of the five individual tables.

*CA of stacked
tables*

Applying CA to the 44×4 matrix of stacked tables results in the map of Exhibit 17.2. The relative positions of the four reponses, *W*, *w*, *H* and *?*, appear almost the same as in Exhibit 16.8. Compared to Exhibits 16.5 and 16.7 the positions are slightly rotated (rotations are discussed in the Epilogue). Each demographic category is defined by a profile of responses and finds its position in the map relative to the four response categories. The following features of the map are of special interest:

• The categories of an ordinal variable such as age can be connected, as shown in the map. Age follows the curved pattern of the responses *W–w–H* from liberal to traditional, as might be expected.

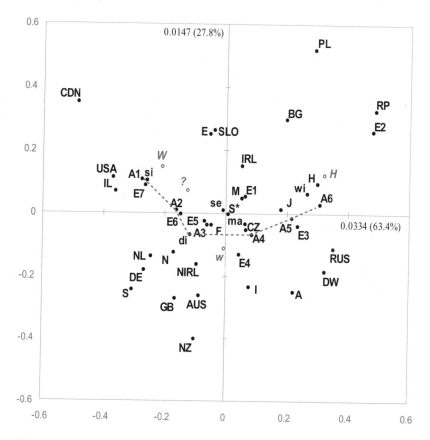

Exhibit 17.2:
*Symmetric CA map
of five stacked
contingency tables
shown schematically
in Exhibit 17.1;
total inertia =
0.05271, percentage
inertia in map:
91.2%.*

- Education has a similar pattern, but from right to left, except for category E1 (no formal education), which is near the average.

- Categories of marital status show si (single) on the liberal side, and wi (widowed) on the traditional side, probably correlated with age group.

- The male and female points M and F lie opposite each other with respect to the average, showing the overall differences between males and females across all countries (we saw the specific differences in Exhibit 16.5).

- Of all the demographical variables, the cross-national differences are still the most important on this issue.

- Countries such as Spain, Slovenia, Ireland and Bulgaria that lie within the arch are polarized countries with higher than average percentages of both *W* (work full-time) and *H* (stay at home) responses.

- The non-response point *?* lies more towards the liberal side of the map; i.e., its profile across the demographics is more similar to *W* than to *H* (in

Chapter 21 we will see that Canada, for example, has a high percentage of non-responses).

Limitations in interpreting analysis of stacked tables

It is important to realize that Exhibit 17.2 is showing the separate associations between the demographic variables and the question responses, and not the relationships amongst the demographic variables. There is no information in the stacked tables about the relationship between age, education and country; for example, the fact that the youngest age group A1, the highest education group E7, and the countries Canada, USA and Israel all lie on the left-hand side does not mean that these countries have predominantly younger more highly educated respondents. Since the variables are being related separately to the question responses, the interpretation is that the youngest age group, the highest education group and these countries all have a predominant, higher than average percentage of W (work full-time) responses. To confirm any relationships between the demographic variables, cross-tabulations between them need to be made and analysed.

Decomposition of inertia in stacked tables

A very useful result here and in future chapters is the fact that when the same individuals are cross-tabulated and stacked, as in Exhibit 17.1, the total inertia in the stacked CA is the average of the inertias in the individual CAs. This result is illustrated by calculating the inertias in each of the six cross-tabulations shown in Exhibit 17.1:

Table	Inertia
Country	0.14558
Gender	0.00452
Age	0.04216
Marital Status	0.02675
Education	0.04221
Average	*0.05224*

The total inertia of the stacked analysis is 0.05271, slightly higher than the above figure, because there are some missing data for some of the demographics, which introduces some additional inertia into the stacked analysis. The totals in each table vary from 30471 for education (the whole Spanish sample, for example, has education coded as "not available") to 33590 (the full sample) for age and country. The effect of the different totals is to increase the total inertia in the stacked analysis, but only slightly since there are small differences in the column totals of each table. For the above decomposition to hold exactly each table must have the same grand total and thus exactly the same column marginal totals. Looking at the above table of inertias also shows how much more inertia there is between countries than between categories of the other variables; hence the relationship of the question responses with countries must dominate the results.

The idea of stacking can be broadened to include additional questions which are cross-tabulated with demographics. In the ISSP survey from which these data are taken, there were in fact four questions relating to attitudes about women working, each with the same set of four responses: work full-time, work part-time, stay at home and a category gathering the various non-responses. The respondents were asked about women working or not when they were (1) married with no children, (2) with a pre-school child at home, (3) with a schoolchild living at home (the question we have been analysing up to now), and (4) when all children are no longer living at home. Each of the five demographic variables can be cross-tabulated with each of these four questions, leading to 20 contingency tables which can be stacked row- and columnwise as shown schematically in Exhibit 17.3.

Stacking tables row- and columnwise

Questions on working women

| 1 | 2 | 3 | 4 |
| W w H ? | W w H ? | W w H ? | W w H ? |

country (24)

gender (2)

age (6)

marital status (5)

education (7)

Exhibit 17.3: *Stacking of contingency tables which separately cross-tabulate five demographic variables with the responses to the question on women working (W=full-time, w=part-time, H=stay at home, ?=don't know/ non-response).*

Applying CA to the 20 tables stacked in five rows and four columns leads to the map in Exhibit 17.4 where each category is represented by a point. The following features of the map are of special interest:

CA of row- and columnwise stacked tables

- The 16 column categories form an even clearer arch pattern, stretching from *2W* and *3W* at top left down to *3w*, *4w* and *2H* at the bottom and up to *4H* and *1H* at top right. This is a typical result in CA where there is what ecologists call a *gradient* in the data, the gradient here being the liberal to traditional spread of attitudes. More or less one can order the categories along this curved gradient as follows (omitting the non-responses (*?*) from the discussion for the moment):

$$2W–3W–2w\ \&\ 4W–1W–3w–4w–2H–1w–3H–4H–1H$$

which shows how the categories line up from extreme liberal on the left (women should work full-time even when they have children at home) to extreme traditional on right (women should stay at home even though there are no children at home).

Exhibit 17.4:
*Symmetric CA map
of 20 stacked
contingency tables
shown schematically
in Exhibit 17.3;
total inertia =
0.04273, percentage
inertia in map:
71.1%.*

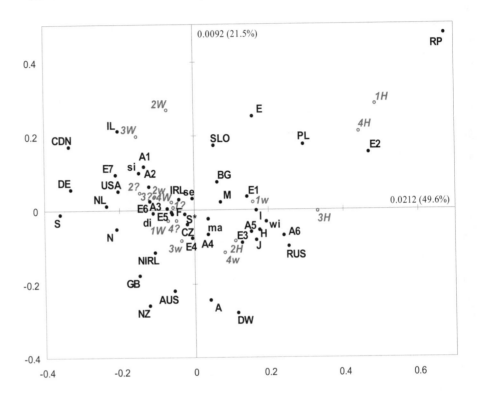

- Most of the demographic points lie along this curve, but there is a substantial spread along the second dimension which opposes groups with a polarized opinion (upper part of map, especially Spain) with groups that have a majority in the intermediate categories of the gradient (lower part of the map, for example Austria and West Germany).

- The four non-response points are all in a small bunch just left of the average — in fact, these points are better represented on the third dimension of this analysis; in other words they should be imagined coming out of the page towards you, which means that the third dimension is mostly a dimension which will line up the demographic groups in terms of their non-response percentages over the four questions.

Variable	Qu. 1	Qu. 2	Qu. 3	Qu. 4	Average
Country	0.15268	0.12834	0.14558	0.13410	*0.14018*
Gender	0.00821	0.00336	0.00452	0.00484	*0.00523*
Age	0.01033	0.03359	0.04216	0.01266	*0.02469*
Marital Status	0.00529	0.01341	0.02675	0.00869	*0.01354*
Education	0.02306	0.02380	0.04221	0.02430	*0.02834*
Average	*0.03991*	*0.04050*	*0.05224*	*0.03692*	*0.04239*

Exhibit 17.5:
Inertias of 20 contingency tables in the stacked table analysed in Exhibit 17.4; averages of the rows and columns are given as well as the overall average.

Again, the result mentioned previously about decomposition of inertia will apply here. First, the exact result is that if every contingency table in the stacked table is the cross-tabulation of exactly the same number of respondents, then the inertia of the stacked table is the average of the inertias of the individual contingency tables. Let us state this a little more formally, since we will be using this result again in the next chapter. Suppose $\mathbf{N}_{qs}$, $q = 1, \ldots, Q$, $s = 1, \ldots, S$ are contingency tables cross-tabulating Q categorical variables pairwise with another set of S categorical variables for the same n individuals (in our example, $Q = 5$ and $S = 4$). Let $\mathbf{N}$ be the stacked table formed by stacking row- and column-wise the $Q \times S$ tables. Then:

Partitioning of the inertia over all subtables

$$\text{inertia}(\mathbf{N}) = \frac{1}{QS} \sum_{q=1}^{Q} \sum_{s=1}^{S} \text{inertia}(\mathbf{N}_{qs}) \qquad (17.1)$$

This result holds approximately if there is a loss of data in some of the contingency tables owing to missing values. In the present example we have combined the missing values for the four questions about women working with other responses such as "don't know" in their respective *?* categories, but there are a few missing values for the demographic variables as a result of the data collection in different countries. So we do not expect the result (17.1) to hold exactly; in fact, the inertia of the stacked table $\mathbf{N}$ (left-hand side of (17.1)) will increase by a small amount ϵ because of differences between the marginal frequencies, so that (17.1) becomes:

$$\text{inertia}(\mathbf{N}) = \frac{1}{QS} \sum_{q=1}^{Q} \sum_{s=1}^{S} \text{inertia}(\mathbf{N}_{qs}) + \epsilon \qquad (17.2)$$

Exhibit 17.5 reports the inertias of all the contingency tables, as well as row and column averages and overall average: as expected, the total inertia in the stacked analysis is slightly higher (0.04273) than the average of the tables (0.04239), which is a difference of 0.8%. Exhibit 17.6 expresses the inertias in Exhibit 17.5 as permills of 0.04273×20 (the left-hand side of (17.2) multiplied by $QS = 20$) to be able to judge the quantities more easily, just as we did when we interpreted numerical contributions in Chapter 11. This shows that on average the countries account for 65.6% of the inertia in the stacked analysis, followed by education (13.3%) and age (11.6%). On question 3 the inertias

Variable	Qu. 1	Qu. 2	Qu. 3	Qu. 4	Total
Country	179	150	170	157	*656*
Gender	10	4	5	6	*24*
Age	12	39	49	15	*116*
Marital Status	6	16	31	10	*63*
Education	27	28	49	28	*133*
Total	*234*	*237*	*306*	*216*	*992*

are generally higher (30.6% of total inertia) — i.e., there are more differences between the demographic groups — while on question 4 they are generally lower (21.6% of total inertia). The total of 992 for this table shows, as we have already remarked above, that 0.8% of the inertia is accounted for by the small disparities between the margins of the 20 contingency tables owing to missing data, i.e., the contribution of ϵ to (17.2).

*Only
"between"
associations
displayed, not
"within"*

Once again we stress the limits of our interpretation of a map such as Exhibit 17.4. When it comes to the four questions, it should be remembered that we are not analysing the associations *within* this set of questions, but rather the associations *between* them and the demographic variables. Analysing associations within a set of variables is the subject of the next chapter on multiple correspondence analysis.

*SUMMARY:
Stacked Tables*

1. An approach to analysing the responses to several questions and their relationships to demographic variables is to concatenate all the contingency tables that cross-tabulate the two sets of variables and to analyse this *stacked table* by regular CA.

2. The interpretation of the CA map of a stacked table is always made bearing in mind that the information being analysed is the set of pairwise relationships between each question and each demographic variable. There is no specific information being mapped about relationships amongst the questions or amongst the demographics.

3. The analysis of a stacked table can be thought of as a consensus or average map from all the CAs of the individual contingency tables.

4. The total inertia of the stacked table is the average of the inertias of each subtable, when the row margins in each row of subtables and the column margins in each column of subtables are identical (this is true when the same number of individuals are cross-tabulated in each subtable). When there is some loss of individuals in some subtables due to missing data, this result is approximate and the total inertia of the subtable will be slightly higher than the average of the inertias.

Multiple Correspondence Analysis

Up to now we have analysed the association between two categorical variables or between two sets of categorical variables where the row variables are different from the column variables. In this and the two following chapters we turn our attention to the association *within* one set of variables, where we are interested in how strongly and in which way these variables are interrelated. In this chapter we will concentrate on the two classic ways to approach this problem, called *multiple correspondence analysis*, or MCA for short. One way is to think of MCA as the analysis of the whole data set coded in the form of dummy variables, called the *indicator matrix*, while the other way is to think of it as analysing all two-way cross-tabulations amongst the variables, called the *Burt matrix*. These two ways are very closely connected, but suffer from some deficiencies which we will try to correct in the following chapter, Chapter 19, where several improved versions of MCA are presented.

Contents

In this chapter we are concerned with s single set of (more than two) variables, usually in the context of a single phenomenon of interest. For example, the four variables used in Chapter 17, on whether women should work or not, could be such a set of interest, or a set of questions about people's attitudes to science, or a set of categorical variables describing environmental conditions at several terrestrial locations. The point is that the set of variables is "homogeneous" in that the variables are of the same substantive type; that is, there is no mix of attitudinal and demographic variables, for example.

A single set of "homogeneous" categorical variables

As an example let us consider the same set of four variables analysed in Chapter 17. The explanation is simplified by avoiding all the cross-cultural differences seen in previous analyses, using only the data from Germany, but

Indicator matrix

Exhibit 18.1:
Raw data and the indicator (dummy variable) coding, for the first six respondents out of N = 3418.

Questions				Qu. 1				Qu. 2				Qu. 3				Qu. 4			
1	*2*	*3*	*4*	*W*	*w*	*H*	*?*	*W*	*w*	*H*	*?*	*W*	*w*	*H*	*?*	*W*	*w*	*H*	*?*
1	3	2	2	1	0	0	0	0	0	1	0	0	1	0	0	0	1	0	0
2	3	3	2	0	1	0	0	0	0	1	0	0	0	1	0	0	1	0	0
4	3	3	2	0	0	0	1	0	0	1	0	0	0	1	0	0	1	0	0
4	4	4	4	0	0	0	1	0	0	0	1	0	0	0	1	0	0	0	1
4	4	4	4	0	0	0	1	0	0	0	1	0	0	0	1	0	0	0	1
1	3	2	1	1	0	0	0	0	0	1	0	0	1	0	0	1	0	0	0
⋮	⋮	⋮	⋮		⋮				⋮				⋮				⋮		

... and so on for 3418 rows

including both the West and East German samples, totalling 3418 respondents (three cases with some missing demographic information were omitted from the original samples — see Computational Appendix, page 235). For the moment we are focusing on the four questions about women working, labelled 1 to 4, each of which has four categories of response, labelled as before: *W* (work full-time), *w* (work part-time), *H* (stay at home) and *?* (don't know/non-response). The *indicator matrix* is the 3418×16 matrix which codes all responses as dummy variables, where the 16 columns correspond to the 16 possible response categories. Exhibit 18.1 illustrates this coding for the first six rows: for example, the first respondent has responses 1, 3, 2 and 2 to the four questions, which are then coded as 1 0 0 0 indicating the response 1 (*W*) to question 1, 0 0 1 0 indicating the response 3 (*H*) to question 2, and 0 1 0 0 indicating the response 2 (*w*) to both questions 3 and 4.

MCA definition number 1: CA of the indicator matrix

The most common definition of MCA is that it is simple CA applied to this indicator matrix. This would provide coordinates for all 3418 rows and 16 columns, but it is mainly the positions of the 16 category points that are of interest for the moment, shown in Exhibit 18.2. The first principal axis shows all four non-response categories together, opposing all the substantive responses. In the previous analysis of these questions (see Exhibit 17.4) where the responses were related to demographic variables, the non-response points were not prominent on the first two axes. But here, because we are looking at relationships within the four questions, this is the most important feature: people who do not respond to one question tend to do the same for the others — for example, amongst the first six respondents in Exhibit 18.1 there are already two respondents who have non-responses for all four questions. On the second axis of Exhibit 18.2, we have the line-up of substantive categories from traditional attitudes at the bottom to liberal attitudes on top. Exhibit 18.3 shows the second and third dimensions of the map, which effectively partials out most of the effect of the non-response points, and the positions of the points are now strikingly similar to those in Exhibit 17.4. Notice that the fact that the liberal side of the horizontal dimension is now on the right is of no

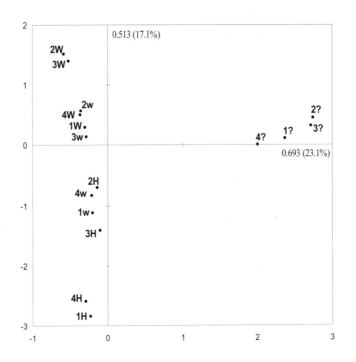

Exhibit 18.2:
MCA map of four questions on women working; total inertia = 3, percentage inertia in map: 40.2%.

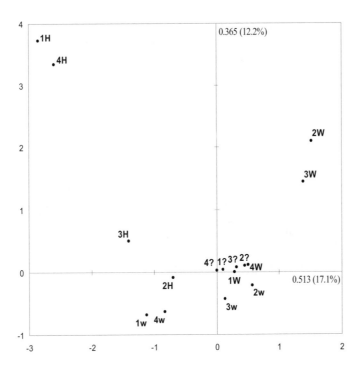

Exhibit 18.3:
MCA map of four questions on women working, showing second and third dimensions; total inertia = 3, percentage inertia in map: 29.3%.

consequence to the interpretation: in fact, it is always possible to reverse an axis (i.e., multiply all coordinates by -1).

Inertia of indicator matrix   The total inertia of an indicator matrix takes on a particularly simple form, depending only on the number of questions and number of response categories and not on the actual data. Suppose there are Q variables, and each variable q has J_q categories, with J denoting the total number of categories: $J = \sum_q J_q$ (in our example, $Q = 4$, $J_q = 4$, $q = 1, \ldots, Q$, and $J = 16$). The indicator matrix, denoted by $\mathbf{Z}$, with J columns, is composed of a set of subtables $\mathbf{Z}_q$ stacked side by side, one for each variable, and the row margins of each subtable are the same, equal to a column of ones. Thus the result (17.1) in Chapter 17 applies: the total inertia of the indicator matrix is equal to the average of the inertias of the subtables. Each subtable $\mathbf{Z}_q$ has a single one in each row, otherwise zeros, so this is an example of a matrix where all the row profiles lie at the vertices, the most extreme association possible between rows and columns; hence the inertias are 1 on each principal axis of the subtable, and the total inertia of subtable $\mathbf{Z}_q$ is equal to its dimensionality, which is $J_q - 1$. Thus the inertia of $\mathbf{Z}$ is the average of the inertias of its subtables:

$$\text{inertia}(\mathbf{Z}) = \frac{1}{Q} \sum_q \text{inertia}(\mathbf{Z}_q) = \frac{1}{Q} \sum_q (J_q - 1) = \frac{J - Q}{Q} \qquad (18.1)$$

Since $J - Q$ is the dimensionality of $\mathbf{Z}$, the average inertia per dimension is $1/Q$. Notice that the first three dimensions that were interpreted in Exhibits 18.2 and 18.3 have principal inertias 0.693, 0.513 and 0.365, all above the average of $1/4 = 0.25$. The value $1/Q$ serves as a threshold for deciding which axes are worth interpreting in MCA (analogous to the threshold of 1 for the eigenvalues in principal component analysis).

Burt matrix   An alternative data structure for MCA is the set of of all two-way cross-tabulations of the set of variables being analysed. The complete set of pairwise cross-tabulations is called the *Burt matrix*, shown in Exhibit 18.4 for the present example. The Burt matrix is a 4×4 block matrix, with 16 subtables. Each of the 12 off-diagonal subtables is a contingency table cross-tabulating the 3418 respondents on a pair of variables. The Burt matrix is symmetric so there are only 6 unique cross-tabulations, which are transposed on either side of the diagonal blocks. The diagonal subtables (by which we mean the tables on the block diagonal) are cross-tabulations of each variable with itself, which is just a diagonal matrix with the marginal frequencies of the variable down the diagonal. For example, the marginal frequencies for question 1 are 2501 *W* responses, 476 *w*s, 79 *H*s and 362 *?*s. The Burt matrix, denoted by $\mathbf{B}$, is simply related to the indicator matrix $\mathbf{Z}$ as follows:

$$\mathbf{B} = \mathbf{Z}^\mathsf{T}\mathbf{Z} \qquad (18.2)$$

1W	1w	1H	1?	2W	2w	2H	2?	3W	3w	3H	3?	4W	4w	4H	4?
2501	0	0	0	172	1107	1131	91	355	1710	345	91	1766	538	40	157
0	476	0	0	7	129	335	5	16	261	181	18	128	293	17	38
0	0	79	0	1	6	72	0	1	17	61	0	14	21	38	6
0	0	0	362	1	57	108	196	7	96	55	204	51	45	2	264
172	7	1	1	181	0	0	0	127	48	4	2	165	15	0	1
1107	129	6	57	0	1299	0	0	219	997	61	22	972	239	13	75
1131	335	72	108	0	0	1646	0	24	989	573	60	760	616	84	186
91	5	0	196	0	0	0	292	9	50	4	229	62	27	0	203
355	16	1	7	127	219	24	9	379	0	0	0	360	14	1	4
1710	261	17	96	48	997	989	50	0	2084	0	0	1348	567	23	146
345	181	61	55	4	61	573	4	0	0	642	0	202	286	73	81
91	18	0	204	2	22	60	229	0	0	0	313	49	30	0	234
1766	128	14	51	165	972	760	62	360	1348	202	49	1959	0	0	0
538	293	21	45	15	239	616	27	14	567	286	30	0	897	0	0
40	17	38	2	0	13	84	0	1	23	73	0	0	0	97	0
157	38	6	264	1	75	186	203	4	146	81	234	0	0	0	465

Exhibit 18.4:
Burt matrix of all two-way cross-tabulations of the four variables of the example on attitudes to women working. Down the diagonal are the the cross-tabulations of each variable with itself.

The other "classic" way of defining MCA is the application of CA to the Burt matrix **B**. Since **B** is a symmetric matrix, the row and column solutions are identical, so only one set of points is shown — see Exhibit 18.5. Because of the direct relationship (18.2), it is no surprise that the solutions are related, in fact at first glance Exhibit 18.5 looks identical to Exhibit 18.2, only the scale has changed slightly on the two axes. This is the only difference between the two analyses — the Burt version of MCA gives principal coordinates which are reduced in scale compared to the indicator version, where the reduction is relatively more on the second axis compared to the first.

MCA definition number 2: CA of the Burt matrix

The two ways of defining MCA are related as follows:

Comparison of MCA based on indicator and Burt matrices

- In both analyses the standard coordinates of the category points are identical — this is a direct result of the relationship (18.2).

- Also as a result of (18.2), the principal inertias of the Burt analysis are the squares of those of the indicator matrix.

- Since the principal inertias are less than 1, squaring them makes them smaller in value (and the lower principal inertias relatively smaller still). The principal coordinates are the standard coordinates multiplied by the square roots of the principal inertias, which accounts for the reduction in scale in Exhibit 18.5 compared to Exhibit 18.2.

- The percentages of inertia are thus always going to be higher in the Burt analysis.

The subtables of the Burt matrix have the same row margins in each set of horizontal tables and the same column margins in each set of vertical tables,

Inertia of the Burt matrix

Exhibit 18.5:
*MCA map of Burt
matrix of four
questions on women
working, showing
first and second
dimensions; total
inertia = 1.145,
percentage inertia in
map: 65.0%.*

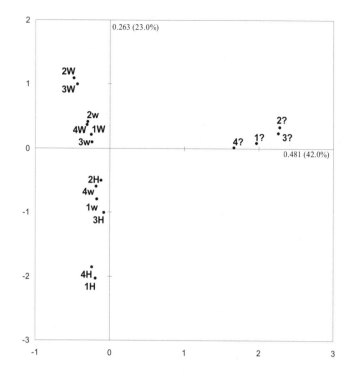

Exhibit 18.5:
*MCA map of Burt
matrix of four
questions on women
working, showing
first and second
dimensions; total
inertia = 1.145,
percentage inertia in
map: 65.0%.*

so the result (17.1) applies exactly: the inertia of **B** will be the average of the inertias of the subtables $\mathbf{B}_{qs}$. Exhibit 18.6 shows the 16 individual inertias of the Burt matrix, and their row and column averages. The overall average is equal to the total inertia 1.145 of **B**. In this table the inertias of the diagonal blocks are exactly 3; in fact their inertias have the same definition (18.1) as the inertias of the subtables of the indicator matrix — they are $J_q \times J_q$ tables of dimensionality $J_q - 1$ with perfect row–column association, and so have maximal inertia equal to the number of dimensions. These high values on the diagonal of Exhibit 18.6 demonstrate why the total inertia of the Burt matrix is so high, which is the cause of the low percentages of inertia on the axes. We return to this topic in the next chapter.

*Positioning
supplementary
variables in the
map*

Suppose we wish to relate the demographic variables gender, age, etc., to the patterns of association revealed in the MCA maps. There are two ways of doing this, highly related, but one of these has some advantages. The first way is to code these as additional dummy variables and add them as supplementary columns of the indicator matrix. The second way is to cross-tabulate the demographics with the four questions, as we did in the stacked analysis of

QUESTIONS	Qu. 1	Qu. 2	Qu. 3	Qu. 4	Average
Qu. 1	3.0000	0.3657	0.4262	0.6457	1.1094
Qu. 2	0.3657	3.0000	0.8942	0.3477	1.1519
Qu. 3	0.4262	0.8942	3.0000	0.4823	1.2007
Qu. 4	0.6457	0.3477	0.4823	3.0000	1.1189
Average	1.1094	1.1519	1.2007	1.1189	1.1452

Exhibit 18.6:
Inertias of each of the 16 subtables of the Burt matrix, from their individual CAs.

Chapter 17, and add these cross-tables as supplementary rows of the indicator matrix or as supplementary rows (or columns) of the Burt matrix. The second strategy is the preferred strategy because it can be used in both forms of MCA as well as in the improved versions that we present in the next chapter. Moreover, it gives the same positions of the supplementary points in both MCA versions and has the same interpretation as the average positions of those cases belonging to the particular demographic category. Exhibit 18.7 shows the positions of five of the demographic variables we used previously, which can be superimposed on the maps of Exhibit 18.2 or 18.5.

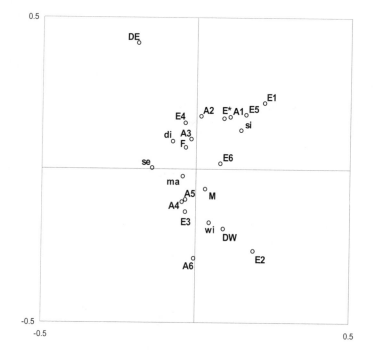

Exhibit 18.7:
Supplementary variables with respect to first two principal axes, to be superimposed on the maps of Exhibits 18.2 or 18.5. These points occupy a small area of the map (note the scale), but will be more spread out in the map of the Burt matrix than that of the indicator matrix.

Based on the positions of the response categories on the first two dimensions of Exhibit 18.2 (similarly, Exhibit 18.5), the farther a demographic category is to the right, the higher will be the frequency of non-responses. The higher up a category is, the more liberal the attitude, and the lower down it is, the more traditional the attitude. Hence West Germany (DW) has more traditional attitudes and more non-responses than East Germany (DE), a pattern that is mimicked almost identically by the male–female (M–F) contrast but not as much as the difference between the two German regions. The age groups show the same trend as before, from young (A1) at the top (liberal) to old (A6) at the bottom (traditional). The lowest education groups have the highest frequency of non-response and the highest education education groups tend to have more liberal attitudes, but so do the lowest education groups E1 and E2. Amongst the marital status groups, single (si) respondents have higher than average non-response and liberal attitudes, opposing separated (se) respondents who have the least non-response, but are otherwise average on the liberal–traditional dimension.

1. MCA is concerned with relationships amongst (or within) a set of variables — usually the variables are homogeneous in the sense that they revolve around one particular issue, and often the response scales are the same.

2. The variables can be recoded as dummy variables in an *indicator matrix*, which has as many rows as cases and as many columns as categories of response. The data in each row are 0s apart from 1s which indicate the particular category of each variable corresponding to the individual case.

3. An alternative coding of such data is as a *Burt matrix*, a square symmetric categories-by-categories matrix formed from all two-way contingency tables of pairs of variables, including on the block diagonal the cross-tabulations of each variable with itself.

4. The two alternative definitions of MCA, applying CA to the indicator matrix or to the Burt matrix, are almost equivalent. Both yield identical standard coordinates for the category points.

5. The difference between the two definitions is in the pricncipal inertias: those of the Burt matrix are the squares of those of the indicator matrix. As a result, the percentages of inertia in the Burt analysis are always more optimistic than those in the indicator analysis.

6. In both approaches, however, the percentages of inertia are artificially low, due to the coding, and underestimate the true quality of the maps as representations of the data.

Joint Correspondence Analysis

Extending simple CA of a two-way table to many variables is not so straight-forward. The usual strategy is to apply CA to the indicator or Burt matrices, but we have seen that the geometry is not so clear any more — for example, the total inertia make little sense and percentages of inertia explained are low. The Burt matrix version of MCA shows that the problem lies in trying to visualize the whole matrix, whereas we are really interested only in the off-diagonal contingency tables which cross-tabulate distinct pairs of variables. Joint correspondence analysis (JCA) concentrates on these tables, ignoring those on the diagonal, resulting in improved measures of total inertia and much better data representation in the maps.

Contents

Exhibit 18.6 shows the inertias in each subtable of the Burt matrix, and their average which is the total inertia of the Burt matrix. This average is inflated by the inertias on the diagonal, which are fixed values equal to the number of categories minus 1 of the corresponding variable ($4 - 1 = 3$ for each variable in that example). Since the analysis tries to explain the inertia in the whole table, the high inertias on the diagonal are going to seriously affect the fit to the whole table. For example, we can evaluate from the results of the MCA how much inertia for each subtable is explained by the two-dimensional MCA map — see Exhibit 19.1. Although the MCA reports that 65.0% of the total inertia is explained, we can see that the off-diagonal tables are explained much better than that, and the tables on the diagonal much worse. By summing the parts explained in the off-diagonal tables and expressing this sum relative to the sum of their total inertias, we find that 83.2% of the off-diagonal tables is explained by the MCA solution (the parts explained in each subtable can

MCA gives bad fit because the total inertia is inflated

QUESTIONS	Qu. 1	Qu. 2	Qu. 3	Qu. 4	Per question
Qu. 1	51.9	78.4	82.5	80.4	61.2
Qu. 2	78.4	55.5	88.2	76.6	65.3
Qu. 3	82.5	88.2	59.6	86.6	69.7
Qu. 4	80.4	76.6	86.6	54.6	63.5

be recovered using the percentages in Exhibit 19.1 and the total inertias in Exhibit 18.6). Similarly, we can calculate that 55.4% of the inertia on the diagonal tables is explained (this is a simple average of the diagonal of Exhibit 19.1 because the total inertias for these tables are constant). So, since we are really interested in only the off-diagonal tables, we should, at least, report a figure such as 83.2% explained rather than 65.0%. But the fit can be made even better than 83.2% as we shall see.

*Ignoring the
diagonal blocks —
joint CA*

It is clear that the inclusion of the tables on the diagonal of the Burt matrix degrade the whole MCA solution. The method is trying to visualize these tables unnecessarily, and moreover these are tables with extremely high inertias, in fact the highest possible inertias attainable. It is possible to improve the calculation of explained inertia by completely ignoring the diagonal blocks in the search for an optimal solution. To do this we need a special algorithm to solve the problem, called *joint correspondence analysis* (JCA). This is an iterative algorithm which performs CA on the Burt matrix in such a way that attention is focused on optimizing the fit to the off-diagonal subtables only. The method starts from the MCA solution and then replaces the diagonal subtables with values estimated from the solution itself, using the reconstitution formula (13.4). Since there is only one set of coordinates and masses for the rows and columns of the symmetric Burt matrix, the formula takes the following form, for a two-dimensional solution, say:

$$\hat{p}_{jj'} = c_j c_{j'} (1 + \sqrt{\lambda_1} \gamma_{j1} \gamma_{j'1} + \sqrt{\lambda_2} \gamma_{j2} \gamma_{j'2}) \qquad (19.1)$$

where $\hat{p}_{jj'}$ is the estimated value of the relative frequency in the (j, j')-th cell of the Burt matrix. Using this formula we replace the diagonal subtables of the Burt matrix with these estimated values, giving a *modified Burt matrix*. CA is then performed on the modified Burt matrix to get a new solution, from which the diagonal subtables are replaced again with estimates from the new solution to get a new modified Burt matrix. This process is repeated several times until convergence, and at each iteration the fit to the off-diagonal subtables is improved.

Results of JCA

Applying JCA to the four-variable data set on women working leads to the following results: 90.2% inertia explained, and percentages for individual tables as shown in Exhibit 19.2. The results are clearly much better than before; all subtables are very well represented, the worst being the cross-tabulation

QUESTIONS	Qu. 1	Qu. 2	Qu. 3	Qu. 4	Per question
Qu. 1	—	97.8	95.8	77.8	88.2
Qu. 2	97.8	—	87.4	97.0	91.8
Qu. 3	95.8	87.4	—	96.7	91.9
Qu. 4	77.8	97.0	96.7	—	88.5

Exhibit 19.2:
Percentage of inertia explained in each of the 12 off-diagonal subtables of the Burt matrix, based on results of JCA of the Burt matrix.

of question 1 with question 4 where the explained inertia is 77.8%. Exhibit 19.3 shows the JCA map, where the scale is intentionally kept identical to that of Exhibit 18.5 for purposes of comparison. It is clear that the solution is practically identical apart from a contraction of the points in scale. In Exhibit 18.5 the principal inertias along the first two axes were 0.481 and 0.263, and here they are 0.353 and 0.128 respectively. So once again there has been a contraction but more of a contraction on the second axis than on the first. This also happened when passing from the MCA of the indicator matrix in Exhibit 18.2 to that of the Burt matrix in Exhibit 18.5, but in that case the standard coordinates of the two analyses were identical — the JCA solution here is different from the MCA solution, as can be seen by closer inspection of Exhibit 19.3 and comparison with the MCA maps in Chapter 18.

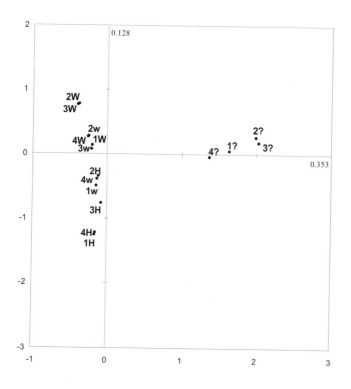

Exhibit 19.3:
JCA map of Burt matrix of four questions on women working; percentage of inertia in map: 90.2%. The percentage of inertia is the sum of the parts explained in each subtable (obtained from Exhibits 19.2 and 18.6) expressed as a percentage of the sum of off-diagonal inertias — see also the Theoretical Appendix, page 207).

JCA results are not nested

The principal axes in JCA are not nested as in the MCA analyses — that is, the solution in two dimensions does not exactly contain the best one-dimensional solution as its first axis, although in practice the nesting is approximate. This is why no percentages of inertia are reported along the axes in Exhibit 19.3 — it is possible to report only a percentage of inertia explained for the solution as a whole, in this case 90.2%. This will affect the reporting of inertia contributions as well: each category point has a certain quality of representation in the map, but we can not break this down into parts for each axis.

Adjusting the results of MCA to fit the off-diagonal tables

The similarity between the JCA solution and the MCA solution occurs in almost all examples in our experience. This suggests that it is mainly the scale change in the solution that distinguishes JCA from MCA; hence, as an alternative, we can investigate simple scale changes of the MCA solution to improve the fit. Given the standard coordinates of the MCA solution, therefore, how should we scale the solution (i.e., define principal coordinates) so that we optimally reconstruct the data in the off-diagonal blocks of the Burt matrix? This is a regression problem, again using the reconstitution formula (19.1), but considering the scale factors (the square roots of the principal inertias) as unknown regression coefficients β_1 and β_2 in the model (illustrated again for a two-dimensional solution):

$$\frac{p_{jj'}}{c_j c_{j'}} - 1 = \beta_1 \gamma_{j1} \gamma_{j'1} + \beta_2 \gamma_{j2} \gamma_{j'2} + e_{jj'} \qquad (19.2)$$

The regression is performed by stringing out all the values on the left-hand side of (19.2) in a vector, just for the cells in the off-diagonal tables, forming the "response variable" — in our four-variable example, with six 4×4 tables off the diagonal, there will be $6 \times 16 = 96$ values in this vector. As "predictor variables" we have the corresponding products $\gamma_{j1}\gamma_{j'1}$ and $\gamma_{j2}\gamma_{j'2}$. A weighted least-squares regression with no constant is performed with weights $c_j c_{j'}$ on the respective values — in our example this gives coefficient estimates $\hat{\beta}_1 = 0.5922$ and $\hat{\beta}_2 = 0.3532$. Squaring these values gives the optimal values 0.351 and 0.125, respectively, for the "principal inertias", for which the explained inertia is 89.9% (this is the coefficient of determination R^2 of the regression). This is the best we can do with the MCA solution — notice how close these are to the principal inertias in the JCA of Exhibit 19.3, which were 0.353 and 0.128. For mapping the categories, the principal coordinates are calculated as the MCA standard coordinates on the first two axes multiplied by the scaling factors $\hat{\beta}_1$ and $\hat{\beta}_2$. But once again the solution is not nested and depends on the dimensionality of the solution — if we perform the same calculation for the three-dimensional solution the first two regression coefficients will not be exactly those obtained above. The nested property will hold only if the "predictors" in (19.2) are uncorrelated. By simply ignoring their correlations, we obtain a simpler (but sub-optimal) way to adjust the solution, which is nested, as described on the following page.

We now describe a simpler adjustment of the principal inertias, which has the nested property; in our experience it gives a solution that is usually very close to optimal. It is also quite easy to compute, involving (i) a recomputation of the total inertia, just for the off-diagonal subtables, and (ii) a simple adjustment of the principal inertias emanating from MCA. Principal coordinates are then calculated in the usual way, as are percentages of inertia.

A simple adjustment of the MCA solution

In MCA of the Burt matrix $\mathbf{B}$, total inertia is the average of the inertias of all subtables, including the offensive ones on the diagonal. But now, as in JCA, the total inertia is the average of the inertias of the off-diagonal subtables. This is easily calculated from the total inertia of $\mathbf{B}$ because we know exactly what the values are of the inertias of the diagonal subtables: $J_q - 1$, where J_q is the number of categories of the q-th variable. Hence

Adjusted inertia = average inertia in off-diagonal blocks

$$\text{sum of inertias of } Q \text{ diagonal subtables} = J - Q \qquad (19.3)$$

while

$$\text{sum of inertias of all two-way tables} = Q^2 \times \text{inertia}(\mathbf{B}) \qquad (19.4)$$

Subtracting (19.3) from (19.4) to obtain the sum of inertias in the off-diagonal blocks and then dividing by $Q(Q-1)$ to obtain the average, leads to:

$$\text{average off-diagonal inertia} = \frac{Q}{Q-1} \times \left(\text{inertia}(\mathbf{B}) - \frac{J-Q}{Q^2} \right) \qquad (19.5)$$

Using our data on women working as an example:

$$\text{average off-diagonal inertia} = \frac{4}{3} \times \left(1.1452 - \frac{12}{16} \right) = 0.5269$$

Another way to compute this value is by directly averaging the inertias of each subtable, which are given in Exhibit 18.6. This needs to be done on only one triangle of the table, since there are only $\frac{1}{2}Q(Q-1) = 6$ pairwise cross-tables:

$$\frac{1}{6}(0.3657 + 0.4262 + 0.6457 + 0.8942 + 0.3477 + 0.4823) = 0.5269$$

Suppose that the principal inertias (eigenvalues) in the MCA of the Burt matrix $\mathbf{B}$ are denoted by λ_k, for $k = 1, 2$, etc. The adjusted principal inertias λ_k^{adj} are calculated as follows:

Adjusting each principal inertia

$$\lambda_k^{\text{adj}} = \left(\frac{Q}{Q-1} \right)^2 \times \left(\sqrt{\lambda_k} - \frac{1}{Q} \right)^2, \quad k = 1, 2, \ldots \qquad (19.6)$$

In our example the first two adjusted inertias are

$$\frac{16}{9} \times (0.6934 - \frac{1}{4})^2 = 0.3495 \text{ and}$$

$$\frac{16}{9} \times (0.5132 - \frac{1}{4})^2 = 0.1232$$

(notice again how close these are to the optimal ones, which were given on the previous page as 0.351 and 0.125).

Adjusted
percentages of
inertia

The adjusted inertias are then expressed relative to the adjusted total to give percentages of inertia along each principal axis:

$$100 \times \frac{0.3495}{0.5269} = 66.3\% \quad \text{and} \quad 100 \times \frac{0.1232}{0.5269} = 23.4\%$$

The percentage of inertia in the two-dimensional adjusted solution is thus 89.7%, only 0.2% less than the optimal adjustment (which is not nested) and 0.5% less than the JCA solution. It has been proved that the percentages calculated according to these simple adjustments give an overall percentage for the solution which is a lower bound for the optimal percentage explained in a JCA solution, as illustrated in this example. Hence, when reporting an MCA, the best way to express measure of fit is as above, and then the square roots of the adjusted principal inertias should be used to scale the standard coordinates to obtain principal coordinates for mapping. We do not give the map here since the relative positions of the points are the same as in Exhibits 18.2 and 18.5 — just the scale is different, more like the scale of Exhibit 19.2.

Data set 10:
News interest in
Europe

As another example of JCA, and also to illustrate supplementary points, consider a large data set from the Eurobarometer survey of 2005 on interest in science. As part of this survey each respondent was asked how interested he or she was in the following six news issues: sports news (S), politics (P), new medical discoveries (M), environmental pollution (E), new inventions and technologies (T), and new scientific discoveries (D). The response scale was "very interested" (++), "moderately interested" (+) and "not at all interested" (0). Hence the response categories are depicted by, for example, E+ for "moderately interested in environmental pollution", or P0 for "not at all interested in politics". In order to avoid the usual phenomenon of non-responses that strongly affect the results, as in the previous example, respondents with any "don't know" and missing responses have been omitted, which reduced the sample size from 33190 to 29652, a reduction of 10.7% — we shall deal with non-response issues specifically in Chapter 21. The adjusted MCA map of these data is shown in Exhibit 19.4. The map shows the "no interest" points forming a diagonal spread of their own to the right and the "interest" points spreading from "moderately interested" at the bottom to "very interested" top left. This is an example of a map that would benefit from a rotation of the axes, if one wanted these two point lines of dispersion to coincide more with the principal axes — see the remark about rotations in CA in the Epilogue. The first axis accounts for 67.0% of the inertia and defines a scale of general interest in news issues. The second axis (22.0%) shows the interest in scientific discovery and technological innovation at the extremes, indicating high correlation between these two. The two points for interest in sports, however, are near the centre of this spread of points, which indicates that high interest in sports (S++), for example, must be also associated with moderate interest in the other issues, and vice versa. Remember that what is being visualized is the association of each category of a particular variable with the categories of all the other variables.

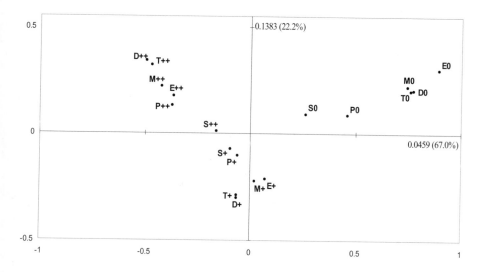

Exhibit 19.4:
*Adjusted MCA map
of news interest
data. Percentage
inertia in map:
89.2%. (If the MCA
of the indicator
matrix were
performed, the
percentage of
explained inertia
would be only
41.1%.)*

Even though we do not show the positions of the 29652 cases in this data set, they can be imagined in the map as supplementary points. That is, if we added the huge 29652 × 18 indicator matrix to the Burt matrix as supplementary rows, each respondent would have a position in the MCA map (but notice that there are only $3^6 = 729$ unique response patterns, so the respondents would pile up at the points representing each response pattern). Since the standard coordinates in the three different versions of MCA (indicator, Burt and adjusted) are the same, the principal coordinate positions of the respondents would be the same in all three. As stated in Chapter 18, the way to show supplementary categories is to add their cross-tabulations with the active variables as supplementary rows of the Burt matrix. For example, in this data set there are samples from 34 European countries. Each respondent from a particular country has a position in the map and the row of frequencies in the cross-tabulation has a profile exactly at the average position of that country's respondent points. Exhibit 19.5 shows the positions of the countries, labelled by their local names — this display should be imagined overlaid on Exhibit 19.4. Of all the countries **TURKIYE** (Turkey) is the most in the "no interest" direction — about 40% of Turkish respondents express no interest on all issues except environmental pollution (22%); whereas **KYPROS** (Cyprus), **ELLADA** (Greece) and **MALTA** seem to be the most interested — for example, Cyprus has the highest percentages of "very interested" responses in issues of environmental pollution (75%), medical discoveries (62%), technological innovation (53%) and scientific discoveries (55%).

*Supplementary
points in adjusted
MCA and JCA*

Exhibit 19.5:
Supplementary country points in the MCA space of the data set of news interest. The original country names, as given in the Eurobarometer, are given.

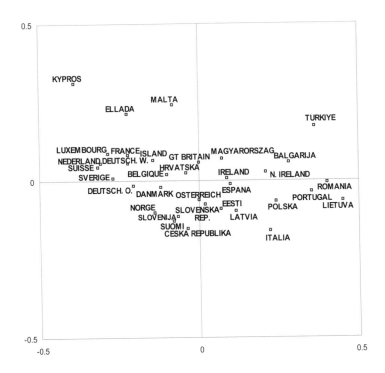

SUMMARY:
Joint
Correspondence
Analysis

1. One way of defining MCA is as the CA of the Burt matrix of all two-way cross-tabulations of a set of variables, including the cross-tables of each variable with itself, which inflate the total inertia.

2. *Joint correspondence analysis* (JCA) finds a map which best explains the cross-tabulations of all pairs of variables, ignoring those on the block diagonal of the Burt matrix. This requires a different iterative algorithm, and results in an optimal solution but one that is not nested.

3. The total inertia to be explained is now the average of all the inertias in the off-diagonal tables of the Burt matrix.

4. An intermediate solution is to condition on the standard coordinates of the MCA solution and find the best weighted least-squares fit to the cross-tables of interest, using regression analysis. However, this solution is again not nested.

5. A simple solution, called *adjusted* MCA, which is nested and thus conserves all the good properties of MCA while solving the low percentage of inertia problem, is to apply certain adjustments to the MCA principal inertias and to the total inertia.

6. Supplementary variables are represented as in all forms of MCA, namely by cross-tabulating them with the active variables and adding them as supplementary rows of the Burt matrix (or modified Burt matrix in JCA).

Scaling Properties of MCA

As was shown in Chapters 7 and 8, there are several alternative definitions of CA and different ways of thinking about the method. In this book we have stressed Benzécri's geometric approach leading to data visualization. In Chapters 18 and 19 it was clear that the passage from simple two-variable CA to the multivariate form of the analysis is not straightforward, especially if one tries to generalize the geometric interpretation. An alternative approach to the multivariate case, which relies on exactly the same mathematics as MCA, is to see the method as a way of quantifying categorical data, generalizing the optimal scaling ideas of Chapter 7. As before, there are several equivalent ways to think about MCA as a scaling technique, and these different approaches enrich our understanding of the method's properties. The optimal scaling approach to MCA is often referred to in the literature as *homogeneity analysis*.

Contents

This data set is taken from the multinational ISSP survey on environment in 1993. We are going to look specifically at $Q = 4$ questions on attitudes towards the role of science. Respondents were asked if they agreed or disagreed with the following statements:

Data set 11: Attitudes to science and environment

A We believe too often in science, and not enough in feelings and faith.

B Overall, modern science does more harm than good.

C Any change humans cause in nature, no matter how scientific, is likely to make things worse.

D Modern science will solve our environmental problems with little change to our way of life.

There were five possible response categories: *1.* strongly agree, *2.* somewhat agree, *3.* neither agree nor disagree, *4.* somewhat disagree, *5.* strongly disagree. For simplicity we have used data for the West German sample only and have omitted cases with missing values on any one of the four questions, leaving us with a sample of $N = 871$ (these data are provided with our **ca** package for R — see the Computational Appendix).

In Chapter 7 CA was defined as the search for quantifications of the categories of the column variable, say, which lead to the greatest possible differentiation, or discrimination, between the categories of the row variable, or vice versa. This is what we would call an "asymmetric" definition because the rows and columns play different roles in the definition and the results reflect this too; for example, the column solution turns out to be in standard coordinates and the row solution in principal coordinates. In Chapter 8 CA was defined "symmetrically" as the search for new scale values which lead to the highest correlation between the row and column variables. Here the rows and columns have an identical role in the definition. These scaling objectives do not include any specific geometric concepts and, in particular, make no mention of a full space in which the data are imagined to lie, which is an important concept in the geometric approach for measuring total inertia and percentages of inertia in lower-dimensional subspaces.

*MCA as a
principal
component
analysis of the
indicator matrix*

The asymmetric definition of optimal scaling, when applied to an indicator matrix, resembles closely principal component analysis (PCA). PCA is usually applied to matrices of continuous-scale data, and has close theoretical and computational links to CA — in fact, one could say that CA is a variant of PCA for categorical data. In the PCA of a data set where the rows are cases and the columns variables (m variables, say, $x_1,...,x_m$), coefficients $\alpha_1,...,\alpha_m$ (to be estimated) are assigned to the columns, leading to linear combinations for the rows (cases) of the form $\alpha_1 x_1 + \cdots + \alpha_m x_m$, called *scores*. The coefficients are then calculated to maximize the variance of the row scores. As before, identification conditions are required to solve the problem, and in PCA this is usually that the sum of squares of the coefficients is 1: $\sum_j \alpha_j^2 = 1$. Applying this idea to the indicator matrix, which consists of zeros and ones only, assigning coefficients $\alpha_1,...,\alpha_J$ to the J dummy variables and then calculating linear combinations for the rows, simply means adding up the α coefficients (i.e., scale values) for each case. Then maximizing the variance for each case sounds just like the optimal scaling procedure of Chapter 7 (maximizing discrimination between the rows); in fact this is almost identical except for one aspect, namely the identification conditions. In optimal scaling the identification conditions would be that the weighted variance (inertia) of the coefficients (not the simple sum of squares) be equal to 1: $\sum_j c_j \alpha_j^2 = 1$. Here the c_j's are the column masses, i.e., the column sums of the indicator matrix divided by the grand total NQ of the indicator matrix — thus each set of c_j's for one categorical variable adds up to $1/Q$. So with this change in the identification condition, MCA could be called the PCA of categorical data, maximizing the variance across cases. The coefficients are the standard coordinates of the column categories, while the MCA principal coordinate of a case is the average of that case's scale values, i.e., $1/Q$ times the sum that was called the score before. The first dimension of MCA maximizes the variance (first principal inertia), the second dimension maximizes the variance subject to the scores being uncorrelated with those on the first dimension, and so on.

MCA as a scaling technique, usually called *homogeneity analysis*, is more commonly seen as a generalization of the correlation approach of Chapter 8. In Equation (8.1) on page 63, an alternative way of optimizing correlation between two categorical variables was given, which is easy to generalize to more than two variables. Again we shall use a pragmatic notation for this particular four-variable example but the ideas easily extend to Q variables which have any number of categories (in our example here, $Q = 4$ and the total number of categories is $J = 20$). Suppose that the four variables have (unknown) scale values a_1 to a_5, b_1 to b_5, c_1 to c_5, and d_1 to d_5. Assigning four of these values a_i, b_j, c_k and d_l to each respondent according to his or her set of responses leads to the quantified responses for the whole sample, which we denote by a, b, c and d (i.e., a denotes all 871 quantified responses to question A, etc.). Each respondent has a score $a_i + b_j + c_k + d_l$ which is the sum of the scale values, so the scores for the whole sample are denoted by $a + b + c + d$. In this context the variables are often referred to as *items* and we talk of the values in a to d as *item scores* and those in $a + b + c + d$ as the *summated scores*. The criterion for finding optimal scale values is thus to maximize the average squared correlation between the item scores and the summated score:

$$\text{average squared correlation} = \frac{1}{4}[\text{cor}^2(a, a + b + c + d) + \text{cor}^2(b, a + b + c + d)$$
$$+ \text{cor}^2(c, a + b + c + d) + \text{cor}^2(d, a + b + c + d)] \quad (20.1)$$

(cf. the two-variable case in (8.1.) on page 63). Again, identification conditions are required, and it is convenient to use the mean 0 and variance 1 conditions on the summated scores: $\text{mean}(a + b + c + d) = 0$, $\text{var}(a + b + c + d) = 1$. The solution to this maximization problem is then given exactly by the standard coordinates of the item categories on the first principal axis of MCA, and the maximized average squared correlation of (20.1) is exactly the first principal inertia (of the indicator matrix version of MCA).

MCA of scientific attitudes example

Exhibit 20.1 shows the two-dimensional MCA map based on the indicator matrix, showing again the very low percentages of inertia (the percentages based on adjusted inertias are 44.9% and 34.2% respectively). But in this case the percentages should be ignored, since it is the values of the principal inertias that are of interest *per se*, being average squared correlations. The maximum value of (20.1) is thus 0.457. The second principal inertia, 0.431, is found by looking for a new set of scale values which lead to a set of scores which are uncorrelated with those obtained previously, and which maximize (20.1) — this maximum is the value 0.431. And so it would continue for solutions on subsequent axes, always uncorrelated with the ones already found. Looking

Exhibit 20.1:
*MCA map
(indicator matrix
version) of science
attitudes, showing
category points in
principal
coordinates. Since
the principal
inertias differ only
slightly (and even
less in their square
roots), the principal
coordinates are
almost the same
contraction of the
standard coordi-
nates on both axes.*

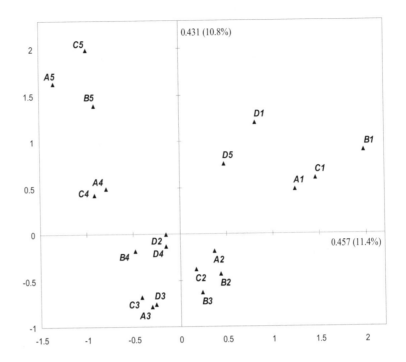

at the map in Exhibit 20.1, we see that questions *A*, *B* and *C* follow a very similar pattern, with strong disagreements on the left to strong agreements on the right, in a wedge-shaped horseshoe pattern. Question *D*, however, has a completely different trajectory, with the two poles of the scale very close together. Now the first three questions were all worded negatively towards science whereas question *D* was worded positively, so we would have expected *D5* to lie towards *A1*, *B1* and *C1*, and *D1* on the side of *A5*, *B5* and *C5*. The fact that *D1* and *D5* lie close together inside the horseshoe means that they are both associated with the extremes of the other three questions — the most likely explanation is that some respondents are misinterpreting the change of direction of the fourth question.

*Individual
squared
correlations*

Knowing the values of the individual squared correlations composing (20.1) will also be interesting information. These can be obtained directly by adding up the individual inertia contributions to the first principal inertia for each question. The results of a MCA usually give these expressed in proportions or permills, so we show these as permills in Exhibit 20.2 as an illustration of how to recover these correlations. Questions *A* to *D* thus contribute proportions 0.279, 0.317, 0.343 and 0.062 of the principal inertia of 0.457. Since 0.457 is the average of the four squared correlations, the squared correlations and thus the correlations are:

Exhibit 20.2:
Permill (‰)
contributions to
inertia of first
principal axis of
MCA (indicator
matrix version) of
data on science and
environment.

| | | QUESTIONS | | | |
CATEGORIES	A	B	C	D	Sum
1 "strongly agree"	115	174	203	25	518
2 "somewhat agree"	28	21	6	3	57
3 "neither–nor"	12	7	22	9	49
4 "somewhat disagree"	69	41	80	3	194
5 "strongly disagree"	55	74	32	22	182
Sum	279	317	343	62	1000

A: $0.279 \times 0.457 \times 4 = 0.510$ correlation $= \sqrt{0.510} = 0.714$

B: $0.317 \times 0.457 \times 4 = 0.579$ correlation $= \sqrt{0.579} = 0.761$

C: $0.343 \times 0.457 \times 4 = 0.627$ correlation $= \sqrt{0.627} = 0.792$

D: $0.062 \times 0.457 \times 4 = 0.113$ correlation $= \sqrt{0.113} = 0.337$

This calculation shows how much lower the correlation is of question D with the total score. Notice that, although the MCA of the indicator matrix was the worst from the usual CA geometric point of view of χ^2-distances, total inertia, etc., the principal inertias and the contributions to the principal inertias do have a very interesting interpretation by themselves. In the approach called *homogeneity analysis*, which is theoretically equivalent to the MCA of the indicator matrix but which interprets the method from a scaling viewpoint, the squared correlations $0.510, 0.579, 0.627$ and 0.113 are called *discrimination measures*.

Loss of homogeneity

In homogeneity analysis the objective function (8.3) (see Chapter 8, page 63) is generalized to many variables. Using the notation above for the present four-variable example, we would calculate the average score $\frac{1}{4}(a_i + b_j + c_k + d_l)$ of the item scores for each respondent and then calculate that respondent's measure of variance within his or her set of quantified responses:

$$\text{variance (for one case)} = \frac{1}{4}(\ [a_i - \frac{1}{4}(a_i + b_j + c_k + d_l)]^2$$
$$+ [b_j - \frac{1}{4}(a_i + b_j + c_k + d_l)]^2$$
$$+ [c_k - \frac{1}{4}(a_i + b_j + c_k + d_l)]^2$$
$$+ [d_l - \frac{1}{4}(a_i + b_j + c_k + d_l)]^2 \ ) \qquad (20.2)$$

The average of all these values over the N cases is then calculated, called the *loss of homogeneity* and the objective is to minimize this loss. Again the MCA (indicator matrix version) solves this problem and the minimized loss is 1 minus the first principal inertia; i.e., $1 - 0.457 = 0.543$. Minimizing the loss is equivalent to maximizing the correlation measure defined previously.

Geometry of
loss function in
homogeneity
analysis

The objective of minimizing loss has a very attractive geometric interpretation which is closely connected to the row-to-column distance definition of CA discussed in Chapter 7. In fact, the homogeneity loss function is exactly the weighted distance function (7.6) on page 55, applied to the indicator matrix. Exhibit 20.3 shows the asymmetric MCA map of all $N = 871$ respondents (in principal coordinates) and the $J = 20$ category points (in standard coordinates), which means that the respondents lie at the centroids of the categories, where the weights are the relative values in the rows of the indicator matrix. Each respondent has a profile consisting of zeros apart from values of $\frac{1}{4}$ in

Exhibit 20.3:
Asymmetric MCA
map (indicator
matrix version) of
science attitudes,
showing respondents
in principal
coordinates and
categories in
standard
coordinates. Each
respondent is at the
average of the four
categories given as
responses. MCA
minimizes the sum
of squared distances
between category
points and
respondents.

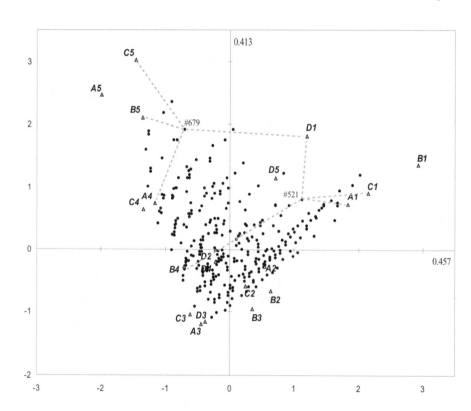

the positions of the four responses; hence each repondent point lies at the ordinary average position of his or her responses. Two respondents, #679 and #521, are labelled in Exhibit 20.3. Respondent #679 chose the categories ($A4,B5,C5,D1$), disagreeing with the first three questions and agreeing to the fourth — those categories are linked to the respondent point on the left-hand side of the display. This is a strong and consistent position in favour of science.

Respondent #521, however, has a mixed opinion: (*A1,B4,C1,D1*), strongly agreeing that we believe too much in science and that human interference in nature will make things worse, but at the same time strongly agreeing that science will solve our environmental problems while disagreeing that science does more harm than good. This shows one of the reasons why *D1* has been pulled to the middle between the two extremes of opinion. Every respondent is at the average of the four categories in his or her set of answers. For any configuration of category points, the respondents could be located at average positions, but the result of Exhibit 20.3 is optimal in the sense that the lines linking the respondents to the category points are the shortest possible (in terms of sum of squared distances). Showing all the links between respondents and their response categories has been called a *star plot*, so the objective of MCA can be seen as obtaining the star plot with the shortest links in the least-squares sense. The number of links between the N respondent points and the corresponding Q category points is NQ, and the value of the loss is actually the average of the squares of the links (for example, in (20.2) where $Q = 4$ the sum of the four squares is divided by 4, and then the average over N is calculated, so that the sum of squared values is divided by $4N$). So the average sum of squared links on the first dimension is $1 - 0.457 = 0.513$, and on the second dimension it is $1 - 0.413 = 0.587$; by Pythagoras' theorem, the average sum of squared links in the two-dimensional map of Exhibit 20.3 is $0.513 + 0.587 = 1.100$.

In the present example of the science and environment data, we saw that the question D is not correlated highly with the others (see page 157). If we were trying to derive an overall measure of attitude towards science in this context, we would say that these results show us that question D has degraded the *reliability* of the total score, and should preferably be removed. In reliability theory, the Q variables, or items, are supposed to be measuring one underlying construct. *Cronbach's alpha* is a standard measure of reliability, defined in general as:

$$\alpha = \frac{Q}{Q-1}\left(1 - \frac{\sum_q s_q^2}{s^2}\right) \qquad (20.3)$$

where s_q^2 is the variance of the q-th item score, $q = 1,\ldots,Q$ (e.g., variances of a, b, c and d) and s^2 is the variance of the average score (e.g., variance of $\frac{1}{4}(a+b+c+d)$). Applying this definition to the first dimension of the MCA solution, it can be shown that Cronbach's alpha reduces to the following:

$$\alpha = \frac{Q}{Q-1}\left(1 - \frac{1}{Q\lambda_1}\right) \qquad (20.4)$$

where λ_1 is the first principal inertia of the indicator matrix. Thus the higher the principal inertia, the higher the reliability. Using $Q = 4$ and $\lambda_1 = 0.4574$ (four significant digits for slightly better accuracy) we obtain:

$$\alpha = \frac{4}{3}\left(1 - \frac{1}{4 \times 0.4574}\right) = 0.605$$

Having seen the behaviour of question D, an option now is to remove it and re-compute the solution with the three questions that are highly intercorrelated. The results are not given here, apart from reporting that the first principal inertia of this three-variable MCA is $\lambda_1 = 0.6018$, with an increase in reliability to $\alpha = 0.669$ (use (20.4) with $Q = 3$). As a final remark, it is interesting to notice that the average squared correlation of a set of random variables, with no zero pairwise correlation between them, is equal to $1/Q$, and this corresponds to a Cronbach's alpha of 0. The value $1/Q$ is exactly the threshold used in (19.7) for adjustment of the principal inertias (eigenvalues), and was also the average principal inertia in the MCA of the indicator matrix, mentioned in Chapter 18.

SUMMARY:
Scaling Properties
of MCA

1. Optimal scaling in a two-variable context was defined as the search for scale values for the categories of one variable which lead to the highest separation of groups defined by the other variable. This problem is equivalent to finding scale values for each set of categories which lead to the highest possible correlation between the row and column variables.

2. In a multivariate context, optimal scaling can be generalized as the search for scale values for the categories of all variables so as to optimize a measure of correlation between the variables and their sum (or average). Specifically, the average squared correlation is maximized between the scaled observations for each variable, called *item scores*, and their sum (or average), called simply the *score*.

3. Equivalently, a minimum can be sought for the variance between item scores within each respondent, averaged over the sample. This is the usual definition of *homogeneity analysis*.

4. The scaling approach in general, exemplified by homogeneity analysis, is a better framework for interpreting the results of MCA of an indicator matrix. The principal inertias and their breakdown into contributions are more readily interpreted as squared correlations, rather than quantities with a geometric significance as in simple CA.

5. The first principal inertia in the indicator matrix version of MCA has a monotonic functional relationship with *Cronbach's alpha* measure of reliability: the higher the principal inertia, the higher the reliability.

6. Since the standard coordinates are identical for the MCA of the indicator matrix, the Burt matrix and in the adjusted form, these scaling properties apply to all three versions of MCA.

Subset Correspondence Analysis

It is often desirable to restrict attention to part of a data matrix, leaving out either some rows or some columns or both. For example, the columns might subdivide naturally into groups and it would be interesting to analyse each group separately. Or there might be categories corresponding to missing values and one would like to exclude these from the analysis. The most obvious approach would be simply to apply CA to the submatrix of interest — however, one or both of the margins of the submatrix would differ from those of the original data matrix, and so the profiles, masses and distances would change accordingly. The approach presented in this chapter, called *subset correspondence analysis*, fixes the original margins of the whole matrix, using these to determine the masses and χ^2-distance in the analysis of any submatrix. Subset CA has many advantages; for example, the total inertia of the original data matrix is decomposed amongst the subsets, hence the information in a data matrix can be partitioned and investigated separately.

Contents

Analysing the consonants and vowels of author data set

The author data set of Exhibit 10.6 is a good example of a table with columns that naturally divide into subsets — the 26 letters of the alphabet formed by 21 consonants and 5 vowels. We have seen in Chapter 10, page 79, that the total inertia of this table is very low, 0.01873, but that there is a definite structure amongst the rows (the 12 texts by the six authors). It would be interesting to see how the results are affected if we restrict our attention to the subset of consonants or the subset of vowels. One way of proceeding would be simply to analyse the two submatrices, the 12×21 matrix of consonant frequencies and the 12×5 matrix of vowel frequencies. But this means that the values in the profiles of each text would be recalculated with respect to the new margins of the submatrix being analysed. In the analysis of consonants, for example, the relative frequencies of *b*, *c*, *d*, *f*, ..., etc. would be calculated relative to the total number of consonants in the text, not the total letters. Moreover, the mass of each text would be proportional to the number of consonants counted, not the number of letters. As for the consonants, their profiles would remain the same as before but the χ^2-distances between them would be different because they depend on the row masses, which have changed.

Subset analysis keeps original margins fixed

An alternative approach, which has many advantages, is to analyse the submatrix but keep the original margins of the table fixed for all calculations of mass and distance. Algorithmically, this is a very simple modification of any CA program — all that needs to be done is to suppress the calculation of marginal sums "local" to the submatrix selected, maintaining the calculation of these sums using the original complete table. This method is called *subset correspondence analysis*.

Subset CA of consonants, standard biplot

Applying subset CA to the table of consonant frequencies (see pages 242–244), we obtain the map of Exhibit 21.1. Here the standard CA biplot is given rather than the symmetric or asymmetric map (see Chapter 13). The texts are in principal coordinates, so their interpoint distances are approximate χ^2-distances, where the distances are based on that part of the original χ^2-distance function due only to the consonants, dropping the terms due to the vowels. The consonants are in standard coordinates multiplied by the respective square roots of the relative frequency of the consonant (i.e., relative frequency in the set of 26 letters — remember that the marginal sums are always those of the original table). The squared lengths of the consonant vectors on each axis are proportional to their contribution to the axis, which is why the letter *y* is so prominent on the second axis (more than 50% in this case). This biplot works just as well for tables with low or high inertias and is particularly useful in this example where the inertia is extremely small. Comparing this map with the asymmetric map of Exhibit 10.7, we can see that the letters are pointing in more or less the same directions and that the configuration of the texts is quite similar. The total inertia is 0.01637, and this value is exactly the sum of the inertias of the consonants in the full analysis. On page 79 we reported the total inertia of the full table to be 0.01873; hence the consonants are responsible for 87.4% (0.01637 relative to 0.01873) of the inertia. Having realized that the consonants contribute the major part of the inertia, it is no surprise that most of the structure displayed in the full analysis of Exhibit 10.7 and the subset analysis of Exhibit 21.1 is the same.

Subset CA of the vowels, standard biplot

The total inertia of the orginal table is decomposed as follows between the consonants and the vowels:

$$\text{total inertia} = \text{inertia of consonants} + \text{inertia of vowels}$$
$$0.01873 = 0.01637 + 0.00236$$
$$(87.4\%) \quad (12.6\%)$$

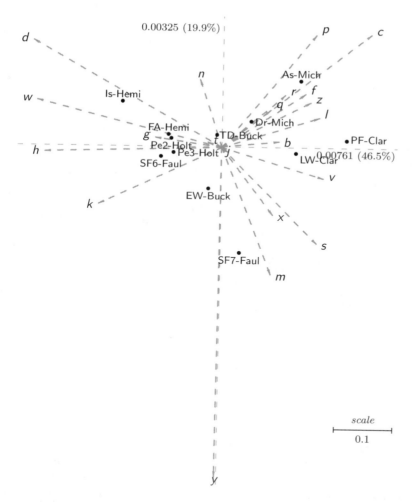

Exhibit 21.1:
Subset analysis of consonants in author example; standard CA row biplot, i.e., rows (texts) in principal coordinates and columns (letters) in standard coordinates multiplied by the square roots of column masses.

The inertia in the vowels part of the table is much smaller, only 12.6% of the original total, in spite of the fact that the vowels are relatively more frequent (38.3% of the letters correspond to the 5 vowels, compared to 61.7% for the 21 consonants). The subset CA of the vowels, again with standard biplot scaling, is shown in Exhibit 21.2, and the lower dispersion of the texts compared to the vectors for the letters is immediately apparent, compared to Exhibit 21.1. However, some pairs of texts are still lying in fairly close proximity. There is an opposition of the letter *e* on the left versus the letter *o* on the right, with a corresponding opposition of the texts by Buck versus those by Faulkner. Of the six authors, the texts of Holt seem to be the most different. In Chapter 25 we shall discuss permutation tests for testing the significance of these results; anticipating this, it turns out the pairing of the texts is highly significant in the case of the consonants but not significant for the vowels.

Exhibit 21.2:
*Subset analysis of
vowels in author
example; standard
CA row biplot, i.e.,
rows (texts) in
principal
coordinates and
columns (letters) in
standard
coordinates
multiplied by the
square roots of
column masses.*

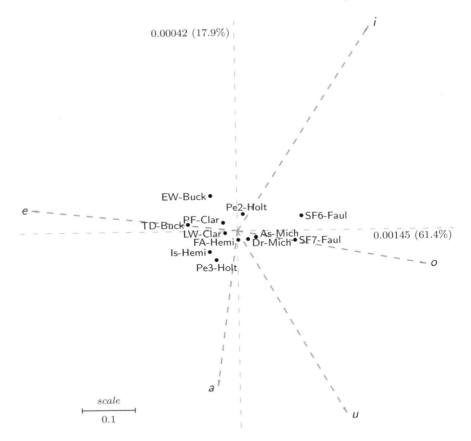

Subset MCA   The subset idea can be applied to MCA in much the same way and provides a very useful tool for investigating patterns in specific categories in multivariate categorical data. In questionnaire surveys it may be interesting from a substantive point of view to focus on a particular subset of responses, for example only the categories of agreement on a five-point agreement–disagreement scale, or all the "middle" response categories ("neither agree nor disagree") or the various non-substantive response categories ("don't know", "no response", "other", etc.). Or we might want to exclude the non-substantive response categories and focus only on substantive answers. In all these cases, the subset analysis will allow us to see more clearly how demographic variables relate to these particular types of responses, which otherwise might not be so clear when all categories are analysed together. The subset option allows us to partition the variation in the data into parts for different sets of categories, which can then be visualized separately. The way to perform subset MCA is to apply subset CA to the appropriate parts of the indicator matrix or Burt matrix, as we illustrate now.

We return to the data set on working women introduced in Chapter 17 and analysed by MCA in Chapter 18. Each of the four questions has a response category for "don't know" and missing responses, previously labelled by the symbol *?* in the maps. These categories were very prominent on the first principal axis of the MCA (see Exhibit 18.2). We can perform a subset analysis on the substantive response categories, labelled *W* (work full-time), *w* (work part-time) and *H* (stay at home) for the four variables, omitting the columns of the indicator matrix corresponding to the non-substantive response *?*, maintaining the original row and column margins of the indicator matrix. Since the row sums of the indicator matrix are 4 in this case, the subset analysis maintains the equal weighting for each row (respondent) and the profile values are still zeros or $\frac{1}{4}$. The respondents with four substantive answers will have four values of $\frac{1}{4}$ in their profiles, while those with three substantive answers will have three $\frac{1}{4}$s and so on. If we simply omitted these columns and performed a regular CA on the indicator matrix, then there would be values of $\frac{1}{3}$ for those with three substantive responses, $\frac{1}{2}$ for those with two, and 1 for those with just one substantive response. The profile of a case with four non-substantive responses would be impossible to calculate, whereas in a subset analysis such a case has a set of zeros as data and is represented at the origin of the map. The total inertia of the subset of 12 categories is 2.1047. Since the total inertia of the whole indicator matrix is 3, this shows that the inertia is decomposed as 2.1047 (70.2%) for the substantive categories and 0.8953 (29.8%) for the non-substantive ones. The principal inertias and percentages of inertia for the first two dimensions of this subset analysis are 0.5133 (24.4% of the total of 2.1047) and 0.3652 (17.4%), i.e., 41.8% in the two-dimensional solution. These percentages again suffer from the problem, as in MCA, of being artificially low. As in Chapter 19, an adjustment of the scaling factors on the axes can be implemented, as will be demonstrated below.

Subset analysis on an indicator matrix

As in regular MCA, the subset analysis can also be performed on the appropriate part of the Burt matrix. To illustrate the procedure, the Burt matrix, given in Chapter 18 in Exhibit 18.4, can be rearranged so that all categories of the subset are in the top left part of the table, as shown in Exhibit 21.3. So the subset of interest is the 12×12 submatrix, itself in a block structure made up of the four sets of three substantive responses, while the four non-substantive categories are now the last rows and columns of the table. The analysis of the subset gives a total inertia of 0.6358 and principal inertias and percentages of 0.2635 (41.4%) and 0.1333 (21.0%) on the first two dimensions: as in MCA, this is an improvement over the indicator matrix version, explaining 62.4% of the inertia compared to 41.8%. Notice that the connection between the indicator and Burt versions of subset MCA is the same as in regular MCA: the principal inertias in the Burt analysis are the squares of those in the indicator version, for example $0.2635 = 0.5133^2$.

Subset analysis on a Burt matrix

Exhibit 21.3:
Burt matrix of four categorical variables of Exhibit 18.4, re-arranged so that all non-substantive response categories (?) are in the last rows and columns. All substantive responses (W, w and H) are in the upper left 12 × 12 part, while the lower right 4 × 4 corner contains the co-occurences of the non-substantive responses ("don't know/missing").

1W	1w	1H	2W	2w	2H	3W	3w	3H	4W	4w	4H	1?	2?	3?	4?
2501	0	0	172	1107	1131	355	1710	345	1766	538	40	0	91	91	157
0	476	0	7	129	335	16	261	181	128	293	17	0	5	18	38
0	0	79	1	6	72	1	17	61	14	21	38	0	0	0	6
172	7	1	181	0	0	127	48	4	165	15	0	1	0	2	1
1107	129	6	0	1299	0	219	997	61	972	239	13	57	0	22	75
1131	335	72	0	0	1646	24	989	573	760	616	84	108	0	60	186
355	16	1	127	219	24	379	0	0	360	14	1	7	9	0	4
1710	261	17	48	997	989	0	2084	0	1348	567	23	96	50	0	146
345	181	61	4	61	573	0	0	642	202	286	73	55	4	0	81
1766	128	14	165	972	760	360	1348	202	1959	0	0	51	62	49	0
538	293	21	15	239	616	14	567	286	0	897	0	45	27	30	0
40	17	38	0	13	84	1	23	73	0	0	97	2	0	0	0
0	0	0	1	57	108	7	96	55	51	45	2	362	196	204	264
91	5	0	0	0	0	9	50	4	62	27	0	196	292	229	203
91	18	0	2	22	60	0	0	0	49	30	0	204	229	313	234
157	38	6	1	75	186	4	146	81	0	0	0	264	203	234	465

Subset MCA with rescaled solution and adjusted inertias

The problem of low inertias is the same here as in MCA: in Exhibit 21.3 there are still 3 × 3 diagonal matrices on the block diagonal of the 12 × 12 submatrix which forms the subset being analysed. As before, it is possible to rescale the solution by regression analysis so that the off-diagonal submatrices are optimally fitted. This involves stringing out the elements of these 6 off-diagonal matrices, each with 9 elements each, as a vector of 54 elements, forming the "y"-variable of the regression. These elements should be expressed as in (19.2), as contingency ratios minus 1. The two "x"-variables (for a two-dimensional solution) are formed by the corresponding products of the standard coordinates. The optimal values for the scale factors are then found by weighted least-squares, as before (see Chapter 19, page 140), giving a fit of $R^2 = 0.849$. Unfortunately, there does not seem to be a simple short cut in this case, as there was for MCA (see (19.5) and (19.7)). Based on the weighted least-squares regression we obtained scaling factors of 0.3570 and 0.1636, which were then used to obtain the principal coordinates and the map in Exhibit 21.4. The squares of these scaling factors are thus the principal inertias, 0.1275 and 0.0268, given on the axes. The percentage of inertia explained by the adjusted two-dimensional solution, i.e., R^2, is 84.9% (strictly speaking, percentages on individual axes cannot be given because the solution is not nested).

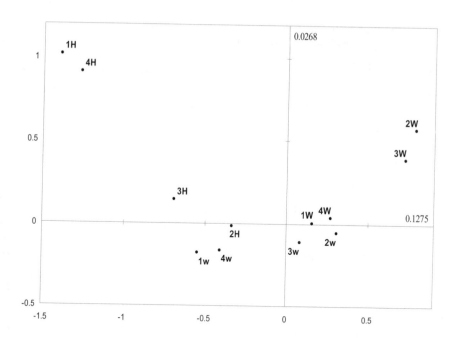

Exhibit 21.4:
Map of subset of response categories, excluding the non-substantive categories. The solution has been adjusted to best fit the off-diagonal tables of the subset matrix, which improves the fit considerably, explaining 84.9% of the inertia.

Displaying supplementary points depends on whether rows or columns have been subsetted. In the case of the author data, for example, where the subset of the vowels was analysed (Exhibit 21.2), the usual centring condition holds for the rows (texts), which were not subsetted, but it does not hold for the columns (vowels). If we wanted to project the letter Y onto the subset map of the vowels, we use the zero-centred row coordinates ϕ_{ik} (i.e., row vertices), so it is not necessary to centre Y's profile, and the usual weighted averaging gives the principal coordinates — see Chapter 12 and the specific transition formula (14.2) applicable to this case (for a two-dimensional solution):

$$\sum_i y_i \phi_{ik} \qquad k = 1, 2 \tag{21.1}$$

Supplementary points in subset CA

where y_i si the ith profile value of Y. On the other hand, to project a new text, with profile values t_j for the subset (these add up to the proportion of vowels in that text, not 1), centring has to be performed with respect to the original centroid values c_j before performing the scalar product operation with the standard column coordinates γ_{jk}:

$$\sum_j (t_j - c_j) \gamma_{jk} \qquad k = 1, 2 \tag{21.2}$$

Notice that to situate a supplementary point in subset CA and in regular CA, this type of centring can always be done, but is not necessary when the standard coordinates satisfy $\sum_i r_i \phi_{ik} = 0$ and $\sum_j c_j \gamma_{jk} = 0$, which is the case when the summation is over the complete set.

Supplementary points in subset MCA

Every respondent (row) of the indicator matrix can be represented as a supplementary point, as well as any grouping of rows into education groups, gender groups, etc. So, as in regular MCA, the categories of supplementary variables are displayed at the centroids of the respondent points that fall into these groups. Exhibit 21.5 shows the positions of various demographic categories with respect to the same principal axes as in Exhibit 12.4.

Exhibit 21.5:
Positions of supplementary points in the map of Exhibit 21.4. Some abbreviations can be found in Chapter 17, page 129; DW and DE are West and East Germany.

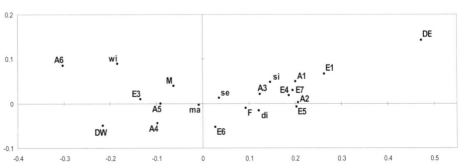

SUMMARY: Subset Correspondence Analysis

1. The idea in *subset CA* is to visualize a subset of the rows or a subset of the columns (or both) in subspaces of the same full space as the original complete set. The original centroid is maintained at the centre of the map, as well as the original masses and χ^2-distance weights.

2. Because the properties of the original space are conserved in the subset analysis, the original total inertia is decomposed exactly into parts of inertia for each subset.

3. Subset CA is implemented simply by suppressing the recomputation of the margins of the subset, and using the original margins (masses) in all the usual CA computations.

4. This idea extends to MCA as well, allowing the selection of any subset of categories, providing an analytic strategy that can be put to great advantage in the analysis of questionnaire data. For example, missing categories can be excluded, or the analysis can focus on one type of response category for all questions, visualizing the dimensions of the subset without any interference from the other categories.

5. As in regular MCA, subset MCA can be applied to the indicator matrix or Burt matrix, and the solutions can be rescaled to optimize the fit to the actual subtables of interest, which dramatically improves the percentages of explained inertia.

6. Supplementary points can also be added to a subset map. In subset MCA this facility allows demographic categories to be related to particular types of response categories.

Analysis of Square Tables

In this chapter we consider the special case when the table of frequencies is square and the rows and the columns refer to the same set of objects in two different states. Such data are found in many situations, for example social mobility tables, confusion matrices in psychology, brand switching tables in marketing research, cross-citations between journals, transition matrices between behavioural states and migration tables. These tables are often characterized by relatively high values down the diagonal, which is such a strong source of association that the more subtle patterns off the diagonal are not seen in the major principal axes. One approach to applying CA to square tables is to split the analysis into two parts: (i) an analysis of the *symmetric* part of the table, which absorbs the main component of inertia, including the diagonal, and (ii) an analysis of the remaining part of the table called the *skew-symmetric* part, which contains the information off the diagonal. It is the visualization of this latter component that shows the amount and direction of "flow" from the rows to the columns and vice versa.

Contents

To give an immediate context to this approach, consider a classic data set on social mobility. This is a historical data set published by Karl Pearson more than 100 years ago on the occupations of fathers and their sons — see Exhibit 22.1. Each father–son pair is counted in one of the cells of the table according to the father's and son's respective occupations. Square tables such as these usually have strong diagonals, since many sons follow their fathers' occupations, but there are some notable asymmetries in the table: for example, in the first line of the table there are 50 fathers in the army, while in the first column there are 84 sons in the army. The flow to the army from other occupations has mostly been from landownership (row 7) and commerce (row 10). Commerce, on the other hand, has had a large outflow to

Data set 12: Social mobility — occupations of fathers and sons

other occupations, with 106 fathers in commerce but only 24 sons, the outflow being mainly to art, divinity, literature and the army.

Exhibit 22.1: *Contingency table between the occupations of fathers and sons. For example, of the 50 fathers employed in the army, 28 of their sons were also in the army, 4 went into teaching/clerical work/civil service, and so on.*

FATHER'S OCCUPATION	*SON'S OCCUPATION* i	ii	iii	iv	v	vi	vii	viii	ix	x	xi	xii	xiii	xiv	*Sums*
Army	28	0	4	0	0	0	1	3	3	0	3	1	5	2	*50*
Art	2	51	1	1	2	0	0	1	2	0	0	0	1	1	*62*
Teaching...*	6	5	7	0	9	1	3	6	4	2	1	1	2	7	*54*
Crafts	0	12	0	6	5	0	0	1	7	1	2	0	0	10	*44*
Divinity	5	5	2	1	54	0	0	6	9	4	12	3	1	13	*115*
Agriculture	0	2	3	0	3	0	0	1	4	1	4	2	1	5	*26*
Landownership	17	1	4	0	14	0	6	11	4	1	3	3	17	7	*88*
Law	3	5	6	0	6	0	2	18	13	1	1	1	8	5	*69*
Literature	0	1	1	0	4	0	0	1	4	0	2	1	1	4	*19*
Commerce	12	16	4	1	15	0	0	5	13	11	6	1	7	15	*106*
Medicine	0	4	2	0	1	0	0	0	3	0	20	0	5	6	*41*
Navy	1	3	1	0	0	0	1	0	1	1	1	6	2	1	*18*
Politics...†	5	0	2	0	3	0	1	8	1	2	2	3	23	1	*51*
Scholarship...°	5	3	0	2	6	0	1	3	1	0	0	1	1	9	*32*
Sums	*84*	*108*	*37*	*11*	*122*	*1*	*15*	*64*	*69*	*24*	*57*	*23*	*74*	*86*	*775*

*Teaching, Clerical Work, Civil Service †Politics & Court °Scholarship & Science

CA of square table

Because this is a contingency table, CA is an appropriate method to visualize it — see Exhibit 22.2. The table has a high inertia (1.297) because of the strong association between rows and columns, so the asymmetric map is used, with father profiles in principal coordinates and son profiles in standard coordinates. If the profile of a father occupation has all zeros except for the value on the diagonal, then that occupation will lie at the vertex of that occupation. The second row for the occupation Art is almost like that, with the highest relative value (51 out of 62, or 82%) of fathers having sons in the same occupation, and this fact is reflected by the separating out of Art in Exhibit 22.2, hence the father-occupation Art almost reaching the son-occupation vertex point *ART*. The row point Crafts is between the vertex points *ART* and *SCH* (*SCHOLARSHIP & SCIENCE*) because there are relatively many sons of fathers in crafts who end up in these two occupations (see row 4 of Exhibit 22.1).

Diagonal of table dominates the CA

The problem with trying to visualize a square matrix such as this one is the presence of the strong diagonal which tends to dominate the analysis. Since CA is trying to explain as much inertia as possible, it is not surprising that the focus is on the high source of inertia on the diagonal, to the detriment of the rest of the table which contains the interesting flows between the occupations.

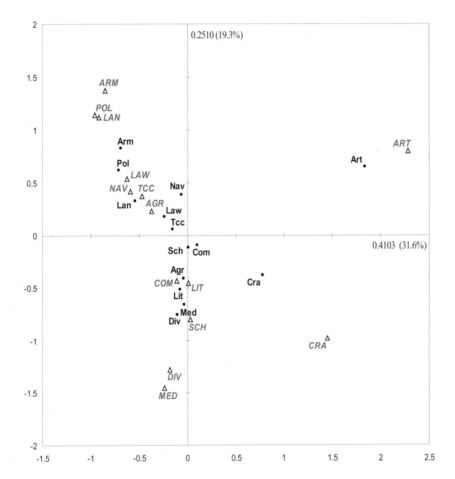

Exhibit 22.2:
*Asymmetric CA
map of mobility
data in Exhibit 22.1,
row points in
principal
coordinates.
Percentage of
explained inertia:
51.0%.*

To back up this assertion with some some figures, the 14 diagonal values
account for 70.9% of the total inertia, while the 182 off-diagonal values account
for 29.1% — i.e., the total inertia is decomposed as follows:

$$\text{total inertia} = \text{inertia on diagonal} + \text{inertia off-diagonal}$$
$$1.2974 = 0.9100 + 0.3774 \tag{22.1}$$
$$100\% = 70.9\% + 29.1\%$$

In the two-dimensional display of Exhibit 22.2, 0.6613 (51.0%) of the total
inertia is explained, and this can also be evaluated for the diagonal and off-
diagonal elements:

$$\text{inertia explained} = \text{inertia explained on diagonal} + \text{inertia explained off-diagonal}$$
$$0.6613 \ (51.0\%) = 0.5483 \ (59.6\%) + 0.1130 \ (29.9\%)$$

The percentages in parentheses are expressed relative to the respective parts of inertias in (22.1), showing that the off-diagonal elements are poorly explained compared to the diagonal ones.

Symmetry and skew-symmetry in a square table It is possible to separate the table into two parts, one part that contains the *symmetric* component of the table, i.e., the average flow between rows and columns, and another part that contains the so-called *skew-symmetric* component quantifying the differential flow. The original table, denoted by $\mathbf{N}$, can be written as follows:

$$\mathbf{N} = \frac{1}{2}(\mathbf{N} + \mathbf{N}^\mathsf{T}) + \frac{1}{2}(\mathbf{N} - \mathbf{N}^\mathsf{T}) \tag{22.2}$$

$$= \mathbf{S} + \mathbf{T}$$

where $\mathbf{S}$ is the symmetric part, containing the averages of elements on opposite sides of the diagonal, and $\mathbf{T}$ the skew-symmetric part, containing half of the differences:

$$s_{ij} = \frac{1}{2}(n_{ij} + n_{ji}) \qquad t_{ij} = \frac{1}{2}(n_{ij} - n_{ji}) \tag{22.3}$$

The following illustrates this decomposition for the top left-hand corner of Exhibit 22.1:

$$
\begin{bmatrix}
28 & 0 & 4 & 0 & \cdots \\
2 & 51 & 1 & 1 & \cdots \\
6 & 5 & 7 & 0 & \cdots \\
0 & 12 & 0 & 6 & \cdots \\
\vdots & \vdots & \vdots & \vdots & \ddots
\end{bmatrix}
=
\begin{bmatrix}
28 & 1 & 5 & 0 & \cdots \\
1 & 51 & 3 & 6.5 & \cdots \\
5 & 3 & 7 & 0 & \cdots \\
0 & 6.5 & 0 & 6 & \cdots \\
\vdots & \vdots & \vdots & \vdots & \ddots
\end{bmatrix}
+
\begin{bmatrix}
0 & -1 & -1 & 0 & \cdots \\
1 & 0 & -2 & -5.5 & \cdots \\
1 & 2 & 0 & 0 & \cdots \\
0 & 5.5 & 0 & 0 & \cdots \\
\vdots & \vdots & \vdots & \vdots & \ddots
\end{bmatrix}
$$

For example, the count of 1 in the second row (father–art) and fourth column (son–crafts) and the count of 12 in the fourth row (father–crafts), second column (son–art), is averaged in $\mathbf{S}$ as 6.5 in both cells, while the deviations (± 5.5) from the average appear in $\mathbf{T}$. The symmetric matrix has the same diagonal as the original table and the property of symmetry: $s_{ij} = s_{ji}$, while the skew-symmetric matrix has zeros on the diagonal and the property of skew-symmetry, namely that elements on opposite sides of the diagonal have the same absolute value but different sign: $t_{ij} = -t_{ji}$.

CA of the symmetric part CA is now applied to the symmetric and skew-symmetric parts separately. Exhibit 22.3 shows the analysis of the symmetric matrix, showing just one set of profile positions because rows and column coordinates are identical. Apart from the single point for each occupation, this map looks very similar to Exhibit 22.2, showing the overall association between the occupations. The first percentage on the axes refers to inertia explained relative to the original asymmetric table, while the percentage in italics refers to inertia explained relative to the total inertia of the symmetric part $\mathbf{S}$ visualized here. Notice that the row and column margins of $\mathbf{S}$ are the averages of the row and column margins of the asymmetric matrix $\mathbf{N}$: if the latter's row and column masses

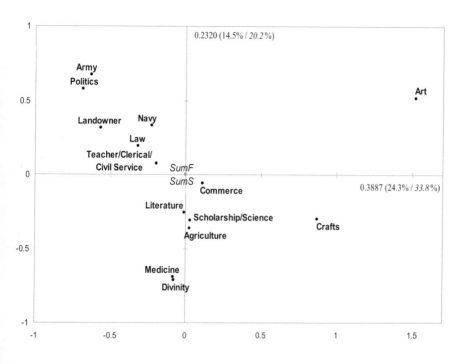

Exhibit 22.3:
CA of symmetric part of Exhibit 22.1. The first percentages are calculated with respect to the total inertia of 1.5991, while percentages in italics are with respect to the inertia of the symmetric part, 1.1485.

are **r** and **c** respectively, then the masses for the rows and columns of **S** are **w** = $\frac{1}{2}$(**r** + **c**).

There are two problems to overcome before CA can be applied to the skew-symmetric matrix **T**. First, **T** has positive and negative values; in fact the sum of the elements of the matrix is zero, and it makes no sense to centre it with respect to its margins, which is the first step in the CA algorithm. The algorithm must be changed so that CA analyses the data without the centring step, just the normalization step which leads to the χ^2-distances. This leads to the second problem: the sums of the rows and columns make no sense as masses. The obvious solution is to adopt the same masses as **S**, i.e., the masses in **w** defined above. This looks like we need a special modified algorithm to analyse **T**, but fortunately there is a result which allows us to obtain the results by a simple recoding of the data.

CA of the skew-symmetric part

This neat recoding trick avoids implementing a special algorithm for calculating the results of the skew-symmetric matrix. The idea is to set up a matrix which is four times the size of the original table **N** in the following format (this is easy to do in R — see page 245 — or in a spreadsheet):

CA of symmetric and skew-symmetric parts in one step

$$\begin{bmatrix} \mathbf{N} & \mathbf{N}^\mathsf{T} \\ \mathbf{N}^\mathsf{T} & \mathbf{N} \end{bmatrix} \tag{22.4}$$

That is, place the transpose of **N** alongside it and below it, and a copy of **N** in the bottom right corner; then apply CA to this block matrix. If **N** is an $I \times I$ matrix, then the block matrix is $2I \times 2I$ and yields $2I - 1$ dimensions, $I - 1$ of which correspond exactly to the dimensions of the symmetric matrix **S** and the remainder to the skew-symmetric matrix **T**. Which dimensions correspond to which parts is easy to see, because the dimensions of the skew-symmetric matrix always occur in pairs of equal principal inertias. In the social mobility example, where $I = 14$, the 27 principal inertias (eigenvalues) are given in Exhibit 22.4. The seven pairs of dimensions with equal principal

Exhibit 22.4:
Principal inertias of all 27 dimensions in the analysis of the 28×28 block matrix (22.4) formed from the social mobility data. The principal inertias that occur in pairs correspond to the skew-symmetric part.

Dim.	Princ. inertia	Dim.	Princ. inertia	Dim.	Princ. inertia
1	0.38868	10	**0.04184**	19	**0.00309**
2	0.23204	11	**0.04184**	20	**0.00309**
3	**0.15836**	12	0.02287	21	0.00166
4	**0.15836**	13	0.02205	22	**0.00115**
5	0.14391	14	**0.01287**	23	**0.00115**
6	0.12376	15	**0.01287**	24	0.00062
7	0.08184	16	0.01036	25	**0.00038**
8	0.07074	17	**0.00759**	26	**0.00038**
9	0.04984	18	**0.00759**	27	0.00015

inertias (shown in boldface), 3 & 4, 10 & 11, 14 & 15, 17 & 18, 19 & 20, 22 & 23, and 25 & 26, correspond to the skew-symmetric analysis, and the other 13 dimensions correspond to the symmetric analysis. The total inertia of the symmetric matrix is the sum of the 13 respective principal inertas: $0.3887 + 0.2320 + 0.1439 + \cdots = 1.1485$, which is 71.8% of the total 1.5991, and the total inertia of the skew-symmetric matrix is the sum of the seven pairs: $2 \times 0.1584 + 2 \times 0.0418 + \cdots = 0.4506$, which is 28.2% of the total (notice that the total inertia, 1.5991, is higher than that of the original matrix, given as 1.2974 in (22.1), because the table is centred at the average margin **w** for both rows and columns).

Visualization of the symmetric and skew-symmetric parts

Dimensions 1 and 2 are thus the best two for visualizing the symmetric matrix: they explain 0.6217 of the inertia of 1.1485, or 54.0%, as shown in Exhibit 22.3. CA of the block matrix yields twice the sets of results for rows and columns, simple repeats of each other, so it is necessary to use only one set of principal coordinates to obtain the map (here we used the first set of coordinates in Exhibit 22.6, dimensions 1 and 2 — see also the Computational Appendix, pages 245–247). Dimensions 3 and 4 are the best for visualizing the skew-symmetric matrix: they explain 0.3167 out of 0.4506, or 70.3% of the inertia of the skew-symmetric part. The map of the skew-symmetric part, shown in Exhibit 22.5, has some unusual properties. First, because of the equality of the principal inertias, the coordinates are free to rotate in the two-dimensional

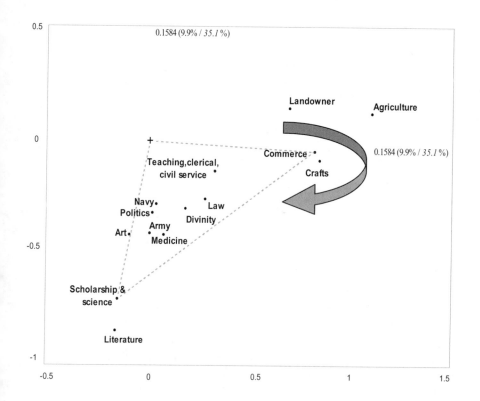

Exhibit 22.5:
CA of skew-symmetric part of Exhibit 22.1. The first percentages are calculated with respect to the total inertia of 1.5991, while percentages in italics are with respect to the inertia of the skew-symmetric part, 0.4506.

map and are not identified with respect to principal axes — hence no axes are drawn in the map. Second, the skew-symmetry of the matrix gives a map where it is necessary to plot only one set of points again (these are again repeated in the CA solution of the block matrix, but with a change of sign — see Exhibit 22.6, where we used the second set of coordinates for dimensions 3 and 4 to make the map in Exhibit 22.5). However, the interpretation is not a distance interpretation but an interpretation of triangular areas in the map. For example, Commerce and Scholarship & Science subtend a large triangle with the origin, which is interpreted as a strong differential flow between these two occupations. The clockwise arrow indicates the direction of the flow from fathers to sons: fathers in Commerce have sons that are going to Scholarship & Science relatively frequently (in Exhibit 22.1, the frequency is 15, whereas there is zero flow in the other direction). Thus, the ocupations Landownership, Agriculture, Commerce and Crafts are experiencing outflows to Literature and Scholarship & Science. Some pairs of occupations make very small triangular areas with the origin, for example Army, Politics and Navy, which means that there is no differential flows between these occupations, but they would be experiencing inflows from Agriculture, Crafts, etc.

OCCUPATION	*Dim. 1*	*Dim. 2*	*Dim. 3*	*Dim. 4*	⋯
Army	-0.632	0.671	-0.011	0.416	⋯
Art	1.521	0.520	0.089	0.423	⋯
Teaching...	-0.195	0.073	-0.331	0.141	⋯
Crafts	0.867	-0.298	-0.847	0.092	⋯
Divinity	-0.077	-0.709	-0.189	0.305	⋯
⋮	⋮	⋮	⋮	⋮	
Army	-0.632	0.671	0.011	-0.416	⋯
Art	1.521	0.520	-0.089	-0.423	⋯
Teaching...	-0.195	0.073	0.331	-0.141	⋯
Crafts	0.867	-0.298	0.847	-0.092	⋯
Divinity	-0.077	-0.709	0.189	-0.305	⋯
⋮	⋮	⋮	⋮	⋮	⋯

SUMMARY:
Analysis of Square
Tables

1. Square tables with the same row and column entities are special because their diagonal values play a major role in the analysis, often obscuring the patterns in the table off the diagonal.

2. An alternative to a regular CA is to split the table into two parts: a *symmetric* part and a *skew-symmetric* part, where the latter part — usually of lower inertia than the symmetric part — encapsulates the asymmetries in the table.

3. The symmetric part is analysed in the usual way, while the skew-symmetric part needs a modified CA algorithm which suppresses the centring and normalization of the table with respect to its margins, which have no sense as masses in this case.

4. The masses used for weighting and χ^2-distances in both analyses are the averages of the row and column masses from the original table.

5. Both analyses can be obtained neatly in one single CA of the table and its transpose set up in a block matrix form. The dimensions corresponding to the symmetric part have unique principal inertias and those corresponding to the skew-symmetric part occur in equal pairs.

6. The map of the symmetric part is interpreted in the usual way, showing the overall association between the entities.

7. The map of the skew-symmetric part has a special geometry where the asymmetries in pairs of entities are visualized as the areas of the triangles that they make with the origin, and the direction of the asymmetry is the same for all pairs.

Data Recoding

In the 22 chapters up to now we have dealt exclusively with frequency tables, either a single table (chapters 1–16 and 22) or in sets (chapters 17–21). In this chapter we will look at other types of data and and how they can be recoded, or transformed, in such a way that CA can still be applied as a method of visualization. This strategy is particularly well developed in Benzécri's approach to data analysis, where CA is the central technique and different data types are preprocessed before being analysed. The types of data treated here are ratings, preferences, paired comparisons and data on a continuous scale. In all of these cases the original CA paradigm should be remembered: CA analyses count data, so if we can transform other types of data to counts of some kind, then it is likely that CA will be appropriate. A standard checklist to perform on the recoded data will be to see if the basic concepts of profile, mass and χ^2-distance make sense in the context of the data.

Contents

Rating scales

We have already met a typical rating scale in Chapter 20, the five-point scale of agreement/disagreement used in the example of science and the environment:

☐	☐	☐	☐	☐
strongly agree	*somewhat agree*	*neither agree nor disagree*	*somewhat disagree*	*strongly disagree*

Previously we treated data on this scale as a nominal categorical variable, creating a dummy variable for each category. This was already an example of data recoding, because CA could not be applied to the original data on a 1-to-5 scale — the notion of a profile would make no sense since a set of responses to the four questions [1 1 1 1] (strongly agree to all four statements) and another set [5 5 5 5] (strongly disagree to all four) would have the same profile. Other types of rating scales often found in social surveys and marketing research are:

— 9-point scale (one extra category between points on 5-point scale):

□   □   □   □   □   □   □   □   □
strongly   *somewhat*   *neither agree*   *somewhat*   *strongly*
agree   *agree*   *nor disagree*   *disagree*   *disagree*

— 4-point scale of importance:

□   □   □   □
not   *fairly*   *very*   *extremely*
important   *important*   *important*   *important*

— 7-point semantic differential scale in a customer satisfaction survey

Service   □   □   □   □   □   □   □   *Service*
unfriendly   *friendly*

— continuous rating scale (e.g., 0 to 10 scale)

Very   0 ——————————————————————— 10   *Very*
dissatisfied   *satisfied*

In this last example the respondent can choose any value between 0 and 10, even with decimal points if desired, but we still think of the data as a rating scale and the recoding will be similar for all the above examples. Notice that when the number of scale points is large, it becomes unwieldy to use the dummy variable coding of MCA.

Doubling of The recoding scheme frequently used in CA for ratings data is called *doubling*.
ratings   The idea behind doubling is to redefine each rating scale as a pair of comple-
               mentary scales, one labelled the "positive", or "high", pole of the scale and
the other the "negative", or "low", pole. Before performing the doubling, it is preferable to have rating scales with a lower endpoint of zero, so 1-to-5 and 1-to-7 scales, for example, should first be converted to 0-to-4 and 0-to-6 respectively, simply by subtracting 1. These values define the data assigned to the positive pole of each scale, assuming a high value refers to the substantively positive end of the scale (e.g., high satisfaction, high importance, high agreement). The negative pole of the scale is then defined as M minus the positive pole, where M is the maximum value of the positive pole (4, 6 or 8 for the ratings scales, 10 for the 0-to-10 scale). Actually, in the agreement–disagreement scale on the previous page, the high value refers to high disagreement, so the labels "+" and "−" would be reversed to avoid confusion — or we could just reverse this scale beforehand. The idea is illustrated for the agreement ratings in the science and environment data set of Chapter 20. Exhibit 23.1 shows the first 10 rows of data and their doubled counterparts. For example, the first value for respondent 1 is a 2, subtracting 1 gives the value 1 and its doubled value is 3, hence the values 1 and 3 in the doubled columns for question 1. These columns are labelled A- and A+ because the first column quantifies how much the respondent disagrees and agrees respectively with the first question. Similarly, the original value of 3 for the second question becomes a 2 and a

Questions				Qu. A		Qu. B		Qu. C		Qu. D	
A	B	C	D	A-	A+	B-	B+	C-	C+	D-	D+
2	3	4	3	1	3	2	2	3	1	2	2
3	4	2	3	2	2	3	1	1	3	2	2
2	3	2	4	1	3	2	2	1	3	3	1
2	2	2	2	1	3	1	3	1	3	1	3
3	3	3	3	2	2	2	2	2	2	2	2
⋮	⋮	⋮	⋮		⋮		⋮		⋮		⋮

... and so on for 871 rows

Exhibit 23.1:
Raw data for the variables on science and environment, and the doubled coding, for the first five respondents out of $N = 871$ (West German sample).

doubled value of 2, i.e., equal values for the disagreement and agreement poles B- and B+, and so on.

The counting paradigm

The doubled values can be thought of as counts in the following sense. The doubled values 1 and 3 are counts of how many scale points are below and above the observed value of 1. The response of 2 ("agree") has one scale point below it (1) and three above it (3, 4 and 5). Similarly, the "neither agree nor disagree" response of 3 is in the middle of the scale and has two scale points below and above it. In this way the doubled data table substitutes the original data by measuring association between each respondent and the agreement and disagreement poles of the rating scale. It is necessary to measure association with both poles for the reason mentioned before: if we use the counts for only one of the poles, then the profiles in CA will make no sense, for example strong agreement to all four questions and strong disagreement would have the same profiles and hence the same position in the map.

CA map of doubled ratings

CA is applied to the doubled table on the right-hand side of Exhibit 23.1, which has 871 rows and 8 columns. The rows all have the same sums (16 in this example); hence the respondent (row) masses are equal, which is correct — there should be no differential weighting of the respondents. Each of the four pairs of columns has the same sum; that is there are four linear restrictions on the columns and not just one as in regular CA. Hence the dimensionality of the data matrix is $8 - 4 = 4$. The total inertia and the decomposition along the four principal axes are as follows:

$$0.3462 = 0.1517 \ (43.8\%) + 0.0928 \ (26.8\%) + 0.0529 \ (15.3\%) + 0.0488 \ (14.1\%)$$

Exhibit 23.2 shows the map of the column points in principal coordinates. There are two points for each question, and the positive poles are directly opposite their negative counterparts relative to the origin, as shown by the dashed lines joining the pairs of poles. The fact that question D is out of line with the other three, which we already saw in Chapter 20, is shown clearly here. We would have expected $D-$ on the right and $D+$ on the left but the direction of this variable is practically at right angles to the others.

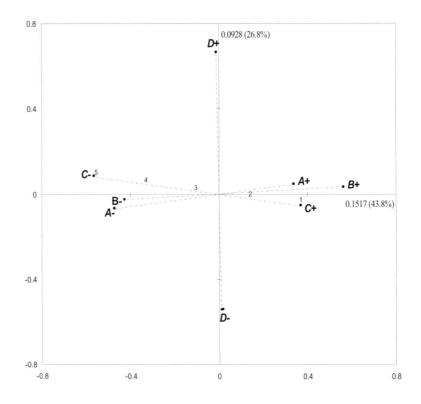

*Rating scale
axes have mean at
origin*

All four rating scale "axes" pass through the origin of the map. The dashed line between the poles can be subdivided into four equal intervals, and labelled by the five scale points (shown for question *C*, using the original 1-to-5 scale where 1 corresponded to strong agreement). The average rating for each question can then be read at the origin on the respective calibrated axis. Thus the average ratings on questions *A* and *C* are more to the agreement (+) side of the scale (the actual average for question *C* is 2.58), while the averages for *B* and *D* are slightly to the disagreement side. Another way of thinking about this is to imagine the endpoints of each rating scale axis having weights proportional to the average of the values attributed to the respective poles — thus *C*+ is closer to the origin than *C*- because it is "heavier".

*Correlations
approximated by
angle cosines*

The cosines of the angles between the four rating scale axes in Exhibit 23.2 are approximate correlations between the variables. Thus we can deduce that variables *A*, *B* and *C* are positively correlated with one another, but uncorrelated with *D*. The four variables have correlation coefficients as follows:

Questions	A	B	C	D
A	1	0.378	0.357	0.036
B	0.378	1	0.436	0.016
C	0.357	0.436	1	-0.062
D	0.036	0.016	-0.062	1

which agrees with our visual deduction. The correlations are not exactly recovered because this map explains only 70% of the inertia. For example, B
and C should make a smaller angle than A and B, but this would be seen
more accurately only in a three-dimensional view of the rating scale axes.

Each respondent has a profile and position in the map, as in a regular CA.
But, as in MCA of survey data with large samples, the individual positions are
not of interest, but rather positions of groups of respondents as supplementary
points. For example, to represent males and females in the six different age
groups in this data set, the average ratings for these 12 groups are computed
and added as supplementary rows (doubled). Their positions are shown in
Exhibit 23.3. All the female groups are on the left-hand side of the map,
that is the disagreement side of questions A, B and C. Apart from the oldest
male group, the male groups are on the agreement side of these questions, i.e.
critical of science's role in the environment.

*Positions of rows
and supplementary
points*

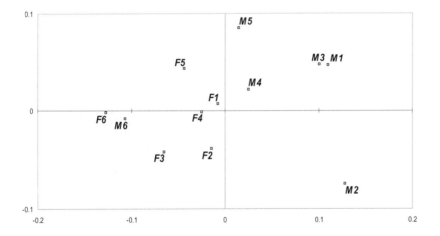

Exhibit 23.3:
*Supplementary
points for males and
females in the six
age groups. The
females are all on
the left-hand side
(disagreement) while
the males — apart
from the oldest
group M6 — are on
the agreement side.*

For purposes of visualizing with CA, preference data may be regarded as a
special case of ratings data. A typical study in marketing research is to ask
respondents to order a set of products from most preferred to least preferred,
or a set of product attributes from most important to least important. As an
example, suppose that there are six products, A to F, and that a respondent
orders them as follows:

Preference data

most preferred : $B > E > A > C > F > D$: least preferred

This ordering corresponds to the following ranks for the six products:

A	B	C	D	E	F
3	1	4	6	2	5

The six ranks are just like ratings on a 6-point scale, the difference being that the respondent has been forced to use each scale point only once. These data can be doubled in the usual way, with the doubled columns assigned labels where + indicates high preference and − low preference:

A−	A+	B−	B+	C−	C+	D−	D+	E−	E+	F−	F+
2	3	0	5	3	2	5	0	1	4	4	1

Frequently, respondents are allowed to rank order a smaller set of most-preferred objects (e.g., first three choices), in which case the objects not ranked are considered to be jointly in the last position which gets the value of a tied rank. For example, if the best three out of six products are rank-ordered, then the three omitted products obtain ranks of 5 each, the average of 4, 5 and 6.

Paired comparisons

Paired (or *pairwise*) *comparisons* are a freer form of preference rankings. For example, each of the 15 possible pairs of the six products A to F is presented to the respondent, who selects the more preferred of the pair. The doubled data for each respondent are then established as follows:

A+: number of times A is preferred to the five other products

A−: number of times the other products are preferred to A $(= 5 − A+)$

and so on. Then proceed as before, applying CA to the doubled data.

Data set 13: European Union indicators

Continuous data can also be visualized with CA after the data is suitably recoded, and several possibilities exist. As an example, consider the data on the left-hand side of Exhibit 23.4, five economic indicators for the 12 European Union countries in the early 1990s. There are a mixture of measurement scales in these data, with unemployment rate and change in personal consumption measured in percentages.

Recoding continuous data by ranks and doubling

A simple recoding scheme is to convert all the observations to ranks, as shown on the right-hand side of Exhibit 23.4. The observations are now ranked within a variable across the countries; for example Luxemburg has the lowest unemployment and so gets rank 1, then Portugal with rank 2 and so on. Tied ranks are given average ranks; for example France and Luxemburg tie for fourth place on the variable *PCP*, so they are given the average 4.5 of ranks 4 and 5. Having transformed to ranks, the doubling can take place as before for each variable: first 1 is subtracted from the ranks to get the negative pole of the scale this time (the low value) and the positive pole is calculated as 11 minus the negative pole. The CA of the doubled matrix is shown in Exhibit 23.5.

COUNTRIES	Original data					Ranked data				
	Unemp	*GDP*	*PCH*	*PCP*	*RULC*	*Unemp*	*GDP*	*PCH*	*PCP*	*RULC*
Belgium	8.8	102	104.9	3.3	89.7	7	7	7	7.5	5.5
Denmark	7.6	134.4	117.1	1	92.4	5	12	11	1	8
Germany	5.4	128.1	126	3	90	3	11	12	6	7
Greece	8.5	37.7	40.5	2	105.6	6	2	2	2	12
Spain	16.5	67.1	68.7	4	86.2	12	4	4	11	3
France	9.1	112.4	110.1	2.8	89.7	8	9	9	4.5	5.5
Ireland	16.2	64	60.1	4.5	81.9	11	3	3	12	2
Italy	10.6	105.8	106	3.8	97.4	10	8	8	10	10
Luxemburg	1.7	119.5	110.7	2.8	95.9	1	10	10	4.5	9
Holland	9.6	99.6	96.7	3.3	86.6	9	6	5	7.5	4
Portugal	5.2	32.6	34.8	3.5	78.3	2	1	1	9	1
UK	6.5	95.3	99.7	2.1	98.9	4	5	6	3	11

Une=Unemployment rate (%) *GDP*=Gross Domestic Product/Head (index)
PCH=Personal Consumption per Head (index) *PCP*=Change in Personal Consumption (%) *RUL*=Real Unit Labour Cost (index)

Exhibit 23.4:
*European Union
economic indicators,
and their ranks from
smallest to largest.*

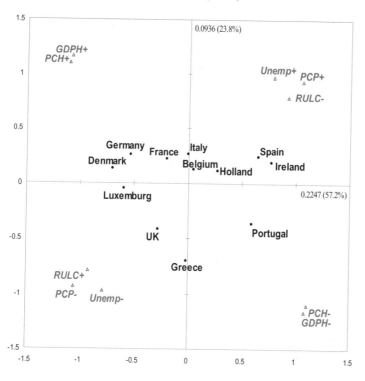

Exhibit 23.5:
*Asymmetric CA
map of European
Union economic
indicators, recoded
as ranks. Inertia
explained is 81.0%.*

Again, the opposite poles of each variable could be connected, but the distances from the origin for each variable are the same in this case because their average ranks are identical (hence plotting just the positive pole would be sufficient). The map shows two sets of variables, strongly correlated within each set but with low correlation between them. Notice that *RULC* (Real Unit Labour Cost) is negatively correlated with *Unemp* (Unemployment Rate) and *PCP* (Percentage Change in Personal Consumption) (when we talk of correlations here, we mean the nonparametric *Spearman rank correlations* because we use the ranks). Each country finds its position in terms of its rank orders on the five variables. Since the ranks are analysed and not the original values, the analysis would be robust with respect to outliers and can be called a *nonparametric* CA of the data.

Other recoding schemes for continuous data

The transformation of the continuous variables to ranks loses some information, although in our experience the loss is minimal in terms of data visualization, and the robustness of the ranks is an advantage in many situations. However, if all the information in the continuous data is needed, other possibilities exist. For example, a transformation that works well is the following: first, convert all variables to standardized values (so-called z-scores) by subtracting their respective means and dividing by standard deviations; then create two doubled versions of each variable from its standardized z using the recoding:

$$\text{positive value} = \frac{1+z}{2} \qquad \text{negative value} = \frac{1-z}{2} \qquad (23.1)$$

Even though it has some negative values, the row and column margins are still positive, and equal for all rows and for all doubled column pairs, so the cases and the variables are weighted equally. CA of this doubled matrix gives a map almost identical to that of Exhibit 23.5. To our knowledge this is the only example of a data matrix with some negative values that can be validly used in CA.

SUMMARY: Data Recoding

1. Data on different measurement scales can be recoded to be suitable for CA.

2. As long as the recoded data matrix has meaningful profiles and marginal sums in the context of the application, CA will give valid visualizations of the data.

3. One of the main recoding ideas is to *double* the variables, that is convert each variable to a pair of variables where the sums of the paired values are constant.

4. Doubling can be performed in the case of ratings, preferences and paired comparisons, leading to a map where each variable is displayed by two points directly on opposite sides of the origin. In the case of ratings data, the origin indicates the average value of the variable on the line connecting its extreme poles.

5. Continuous data can be recoded as doubled ranks, leading to a nonparametric form of CA, or can be transformed to a continous pair of doubled variables, using their standardized values.

Canonical Correspondence Analysis

The objective of CA is to visualize a table of data in a low-dimensional subspace with optimal explanation of inertia. When additional external information is available for the rows or columns, these can be displayed as supplementary points that do not play any role at all in determining the solution (see Chapter 12). By contrast, we may actually want the CA solution to be directly related to these external variables, in an active rather than a passive way. The context where this often occurs is in environmental research, where information on both biological species composition and environmental parameters are available at the same sampling locations. Here the low-dimensional subspace is required which best explains the biological data but with the condition that the space is forced to be related to the environmental data. This adaptation of CA to the situation where the dimensions are assumed to be responses in a regression-like relationship with external variables is called *canonical correspondence analysis*, or CCA for short.

Contents

To motivate the idea behind CCA, we look again at the marine biological data of Exhibit 10.4, page 77. In addition to the species information at each sampling location on the sea bed, several environmental measurements were made: metal concentrations (lead, cadmium, barium, iron, ...), sedimentary composition (clay, sand, pelite, ...) and other chemical measurements such as hydrocarbon and organic content. Since some of these are highly intercorrelated, we chose three representative variables as examples: barium and iron, measured in parts per million, and pelite* as a percentage, shown in Exhibit 24.1. These variables will be used as explanatory variables in a linear regres-

Supplementary continuous variables

---

* Pelite is sediment composed of fine clay-size or mud-size particles.

Exhibit 24.1:
*Environmental data
measured at the 13
sampling points (see
Exhibit 10.4); 11
sites in vicinity of
oil-drilling platform
and 2 reference sites
10km away.*

VARIABLES	S4	S8	S9	S12	S13	S14	S15	S18	S19	S23	S24	R40	R42
Barium (Ba)	1656	1373	3680	2094	2813	4493	6466	1661	3580	2247	2034	40	85
Iron (Fe)	2022	2398	2985	2535	2612	2515	3421	2381	3452	3457	2311	1804	1815
Pelite (PE)	2.9	14.9	3.8	5.3	4.1	9.1	5.3	4.1	7.4	3.1	6.5	2.5	2.0
$\log(Ba)$	3.219	3.138	3.566	3.321	3.449	3.653	3.811	3.220	3.554	3.352	3.308	1.602	1.929
$\log(Fe)$	3.306	3.380	3.475	3.404	3.417	3.401	3.534	3.377	3.538	3.539	3.364	3.256	3.259
$\log(PE)$	0.462	1.173	0.580	0.724	0.623	0.959	0.724	0.613	0.869	0.491	0.813	0.398	0.301

STATIONS (SAMPLES)

sion model within CCA. We prefer to use their values on a logarithmic scale,
a typical transformation to convert ratio-scale measurements on a multiplica-
tive scale to an additive scale — their log-transformed values are also given in
Exhibit 24.1. This transformation not only removes the effect of the different
scales from the three variables but also reduces the influence of large values.

*Representing
explanatory
variables as
supplementary
points*

Before entering the world of CCA, let us first display these three variables on
the map previously shown in Exhibit 10.5. The way to obtain coordinates for
the continuous variables is to perform a weighted least-squares regression of
the variable on the two principal axes, using the column standard coordinates
γ_1 and γ_2 on the first two dimensions as "predictors" and the column masses
as weights, as shown in Chapter 14, page 110. For example, for the regression
of $\log(Ba)$, part of the data are as follows:

Variable	$\log(Ba)$	γ_1	γ_2	Weight
S4	3.219	1.113	0.417	0.0601
S8	3.138	-0.226	-1.327	0.0862
S9	3.566	1.267	0.411	0.0686
⋮	⋮	⋮	⋮	⋮
R42	1.929	2.300	0.7862	0.0326

The results of the regression are:

Source	Coefficient	Standardized coefficient
Intercept	3.322	—
γ_1	-0.301	-0.641
γ_2	-0.229	-0.488

$$R^2 = 0.648$$

The usual way of displaying the variable is to use the standardized regression
ceofficients as coordinates. As illustrated in Chapter 14, page 111, these are

identical to the (weighted) correlation coefficients of $\log(Ba)$ with the two sets of standard coordinates. Repeating the regressions (or, equivalently, calculating the correlation coefficients) the three environmental variables can be placed on the map of Exhibit 10.5, which we show in Exhibit 24.2, omitting the species points. The percentage of variance explained (R^2) for each variable is the sum of the squared correlation coefficients, exactly what we called the *quality* of display of a point. For $\log(Ba)$ it is quite high, 0.648 (or 64.8%) as given above, while for $\log(Fe)$ it is 0.326 and for $\log(PE)$ only 0.126.

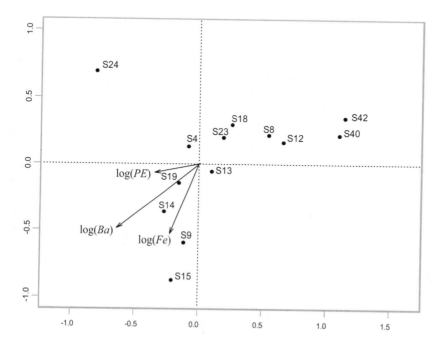

Exhibit 24.2:
Station map of Exhibit 10.5, showing positions of three environmental variables as supplementary points according to their correlations with the two principal axes.

We now turn the problem around: instead of regressing the continuous explanatory variables on the dimensions, we regress the dimensions on the explanatory variables, always incorporating the masses as weights in the regression. The results of the two regression analyses are given in Exhibit 24.3. Notice that the standardized coefficients are, unfortunately, no longer the correlations coefficients we used to display the variables in Exhibit 24.2. For example, the correlations between $\log(Ba)$ and the two dimensions are -0.641 and -0.488, while in the regression analyses above the standardized regression coefficients are -0.918 and -0.327, respectively.

Dimensions as functions of explanatory variables

The percentage of explained variance for the two dimensions (which is actually of inertia since we are weighting the variables by the station masses) is 49.4% and 31.9%, respectively, in the two regressions (see bottom line of Exhibit 24.3). The idea now is to increase this explained inertia by forcing

Constraining the dimensions of CA

Response: CA dimension 1			Response: CA dimension 2		
Source	*Coeff.*	*Stand. coeff.*	*Source*	*Coeff.*	*Stand. coeff.*
Intercept	-9.316	—	Intercept	14.465	—
$\log(Ba)$	-1.953	-0.918	$\log(Ba)$	-0.696	-0.327
$\log(Fe)$	-4.602	0.398	$\log(Fe)$	-3.672	0.318
$\log(PE)$	0.068	0.014	$\log(PE)$	0.588	0.123

$$R^2 = 0.494 \qquad\qquad\qquad R^2 = 0.319$$

the CA solution to be a linear function of the three explanatory variables. The regular CA solution on the marine biological data optimizes the fit to the species profiles with no condition on the dimensions, whereas now we impose the condition that the dimensions be linear combinations of the environmental variables. This will increase the explained inertia of the dimensions as a function of the environmental variables to 100%, but at the same time degrade the explanation of the species data. The way the solution is computed is to project the whole data set onto a subspace which is defined by the three environmental variables, and then perform the CA in the usual way. This is what CCA is: rather than looking for the best-fitting principal axes in the full space of the data, it is the search for principal axes in a constrained, or restricted, part of the space (so CCA could just as well stand for *constrained correspondence analysis*). Having done the CCA (we show the full results later), the regressions of the first two CCA dimensions on the environmental variables are given in Exhibit 24.4. The explained variance (inertia) is now indeed 100%, which is what was intended; by construction, the dimensions are necessarily linear combinations of the environmental variables.

Response: CCA dimension 1			Response: CCA dimension 2		
Source	*Coeff.*	*Stand. coeff.*	*Source*	*Coeff.*	*Stand. coeff.*
Intercept	2.719	—	Intercept	14.465	—
$\log(Ba)$	-2.297	-1.080	$\log(Ba)$	-0.877	-0.412
$\log(Fe)$	1.437	0.124	$\log(Fe)$	12.217	1.058
$\log(PE)$	-0.008	-0.002	$\log(PE)$	-2.378	-0.497

$$R^2 = 1 \qquad\qquad\qquad R^2 = 1$$

CCA restricts the search for the optimal principal axes to a part of the total space, called the *constrained space*, while the rest of the space is called the *unconstrained space* (also called *restricted* and *unrestricted*, or *canonical* and *non-canonical* spaces). Within the constrained space the usual CA algorithm proceeds to find the best dimensions to explain the species data. The search for the best dimensions can also take place within the unconstrained space — this space is the one that is linearly unrelated (uncorrelated) with the environmental variables. So if we are interested in *partialling out* some variables from the analysis, we could do a CCA on these variables and then investigate the dimensions in the unconstrained part of the space.

In the present example, the total inertia of the species-by-sites table is 0.7826 (this is the total inertia of Exhibit 10.4). The inertia in the constrained and unconstrained spaces decompose this inertia into two parts, with values 0.2798 and 0.5028, respectively, i.e., 35.8% and 64.2% of the total inertia. This is an indication why the original CA produced dimensions that were not strongly correlated with the environmental variables because CA tries to explain the maximum inertia possible, and there is more inertia in the unconstrained space than in the constrained one. The decomposition of inertia is illustrated in Exhibit 24.5, including the decomposition along principal axes. Once we constrain the search to the constrained space (depicted by the shaded area in Exhibit 24.5), the first two dimensions have principal inertias of 0.1895 and 0.0615, respectively, totalling 0.2510 or 89.7% of the constrained inertia of 0.2798. Relative to the original total inertia of 0.7826, these are explaining 32.1%, respectively (cf. Exhibit 10.5 where the two-diimensional CA explained 57.5%). On the other hand, in the unconstrained space (not shaded in Exhibit 24.5), if this space is of interest, the first two dimensions have principal inertias 0.1909 and 0.1523, totalling 0.3432 which is 68.3% of the unconstrained inertia 0f 0.5028, or 43.8% of the total inertia. Notice that if the regression of these latter dimensions of the unconstrained space were made on the environmental variables, there would be no relationship (regression coefficients of zero) and explained inertia would be zero.

Decomposition of inertia in CCA

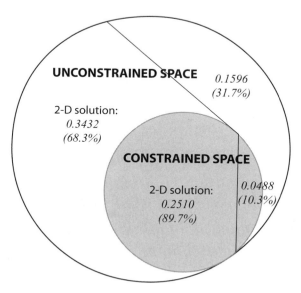

Exhibit 24.5:
Schematic diagram of the decomposition of inertia between the constrained space (shaded) and the unconstrained space, showing the parts of each explained by respective two-dimensional maps. The parts to the right of the straight lines (inertias of 0.0488 and 0.1596) remain unexplained by the respective two-dimensional solutions.

The results of CCA in the constrained space involve coordinates for the usual rows and columns, as in CA, with the same choices for joint plotting, plus the possibility of adding point vectors for the explanatory variables — this is called a *triplot*. The most problematic aspect is how to visualize the explanatory variables: on the one hand, their correlation coefficients with the axes could be used to define their positions, as they would be represented as supplementary points, or their standardized regression coefficients in their relationship to the axes. The latter choice is the preferred one since it reflects the idea that in CCA the axes are linearly related to the explanatory variables. Exhibit 24.6 shows one possible CCA triplot of the present example, where the environmental variables are represented by their regression coefficients. The sites are now in

Exhibit 24.6:
CCA triplot where the species (rows) and sites (columns) are ploted as a row asymmetric map (i.e., sites in standard coordinates), and environmental variables are depicted by their coefficients in their linear relationships with the two axes. The species are shown by symbols in size proportional to their total abundance in the data set, and only a few species' names are indicated for commentary in the text.

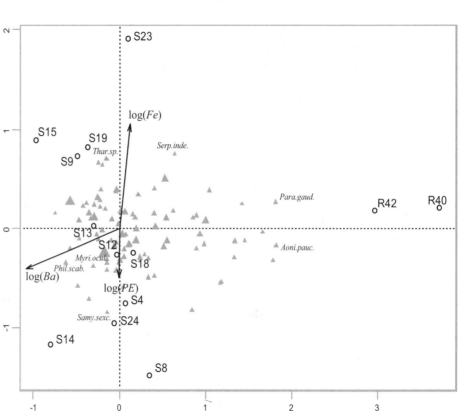

standard coordinates and the species in principal coordinates, so in this sense the basic map is a row principal asymmetric map. As far as the sites and species are concerned, the biplot interpretation holds; since the sites are in standard coordinates they indicate biplot axes onto which each species could be projected to read off its relative abundance at that site (relative to its total abundance across all sites). The site positions along each axis are, by construction, linear combinations of their three standardized values on the environmental variables, using the plotted coefficients. If a site has average value on an environmental variable the contribution of that variable to its position is zero. So the fact that the reference stations R40 and R42 are so far on the other side of log(Ba) means that its values must be low in Barium, which is certainly true. Likewise, S23, S19, S15 and S9 must be high in Iron (especially S23) and S8 and S14 must be high in Pelite. This can be confirmed by looking at the actual figures in Exhibit 24.1. The relationship between the species and the environmental variables is through the sites that they have in common. Species like *"Para.gaud."* and *"Aoni.pauc."* are associated with the reference stations, and these reference stations have low Barium. Species such as *"Thar.sp."* and *"Serp.inde."* are associated with stations that have high Iron and/or low Pelite, while *"Samy.sexc."* down below is associated with stations that have high Pelite and/or low Iron. The reference stations are more or less in the middle of the vertical axis; they are low in both Iron and Pelite, and this has effectively cancelled out their vertical positions.

If there are categorical variables such as *Region* (e.g., with categories Northeast/Northwest/South) or *Rocky* (e.g., with categories yes/no) as explanatory variables, then these are included as dummy variables in the CCA just as they would be included in a regression analysis, that is dropping one of the dummy variables of each set. In the CCA solution these dummy variables are not represented by arrows; rather, the sites that are in each category are averaged (and, as always, applying the usual weights in the averaging), so that each category is represented by a point in the CCA map.

Categorical explanatory variables

An alternative way of thinking about CCA is as an analysis of the weighted averages of the explanatory variables for each species. Exhibit 24.7 shows a small part of this set of averages, for some of the species that have been referred to before. For example, the frequencies of *Myriochele oculata* (*Myri.ocul.*) are given in Exhibit 10.4 as 193, 79, 150, etc., for stations S4, S8, S9, etc., and these stations have values for log(Ba) of 3.219, 3.138, 3.566, etc. So the weighted average for *Myri.ocul.* on that variable is:

Weighted averages of explanatory variables for each species

$$\frac{193 \times 3.219 + 79 \times 3.138 + 150 \times 3.566 + \cdots}{193 + 79 + 150 + \cdots} = 3.393$$

which is the scalar product of the profile of the species with the values of the variable. The "global (weighted) average" in the last line of Exhibit 24.7 is the same calculation using the totals of all species. Hence we can see that *Myri.ocul.* is quite close to the global average, and so does not play as important a role as it did in the CA of Exhibit 10.5. *Para.gaud.* and *Aoni.pauc.* have low averages on log(Ba) because of their relatively high frequencies at the reference sites R40 and R42 where Barium is very low. *Sami.sexc.* has a high average on log(PE) and the reason why *Thar.sp.* and *Serp.inde.* lie at the top is more to do with their low Pelite averages than their high Iron ones.

Exhibit 24.7:
Weighted averages of the three environmental variables across the sites for a selection of species, using the frequencies of the species at each site as weights.

SPECIES	log(Ba)	log(Fe)	log(PE)
		Variables	
Myri.ocul.	3.393	3.416	0.747
⋮	⋮	⋮	⋮
Serp.inde.	3.053	3.437	0.559
Thar.sp.	3.422	3.477	0.651
Para.gaud.	2.491	3.352	0.534
Aoni.pauc.	2.543	3.331	0.537
Samy.sexc.	3.373	3.409	0.971
⋮	⋮	⋮	⋮
Global average	*3.322*	*3.424*	*0.711*

Partial CCA

The idea of partialling out the variation due to some variables can be carried a step further in a *partial CCA*. Suppose that the explanatory variables are divided into two groups, labelled A and B, where the effect of A is not of primary interest, possibly because it is well known, for example a north-south geographical gradient. In a first step the effect of this A set of variables is removed, and in the space not related to these variables a CCA is performed with respect to the set of variables B. There is thus a decomposition of the original total inertia into three parts: the part due to A which is partialled out, and the remainder which decomposes into a part constrained to be related to the B variables and the unconstrained part (which is unrelated to both A and B variables).

SUMMARY: Canonical Correspondence Analysis

1. In CA the dimensions are found which maximize the inertia explained in the solution subspace.

2. In *canonical correspondence analysis* (CCA) the dimensions are found with the same CA objective but with the restriction that the dimensions are linear combinations of a set of explanatory variables.

3. CCA necessarily explains less of the total inertia than CA because it looks for a solution in a constrained space, but it may be this constrained space which is of more interest to the researcher.

4. Total inertia can be decomposed into two parts: the part in the constrained space where the CCA solution is sought, and the part in the unconstrained space which is not linearly related to the explanatory variables. In both these spaces principal axes explaining a maximum amount of inertia can be identified: these are the *constrained* and *unconstrained* solutions respectively.

5. In *partial CCA* the effect of one set of variables is first partialled out before a CCA is performed using another set of explanatory variables.

Aspects of Stability and Inference

Apart from the passing mention of a χ^2 test and the discussion about significant clustering in Chapter 15, this book has concentrated exclusively on the geometric properties of CA and its interpretation. In this final chapter we give an overview of some approaches to investigating the stability of CA solutions and the sampling properties of statistics such as the total inertia, principal inertias and principal coordinates. We make the distinction between (i) stability of the solution, irrespective of the source of the data, (ii) sampling variability, assuming the data arise out of some form of random sampling from a wider population, and (iii) testing specific statistical hypotheses.

Contents

Throughout this book CA has been described as a method of data description, as a way of re-expressing the data in a more accessible graphical format to facilitate the exploration and interpretation of the numerical information. Whether the features in the map are evidence of real phenomena or arise by chance variation is a separate issue. To make statements, or so-called *inferences*, about the population is a separate exercise which requires special consideration, and is feasible only when the data are validly sampled from a wider population. For the type of categorical data considered in this book, there are many frameworks that allow hypotheses to be tested and inferences to be made concerning the characteristics of the population from which the data are sampled. For example, *log-linear modelling* allows interactions between variables to be formally tested for significance, while *association modelling* is closely connected to CA and enables differences between category scale values, for example, to be tested. There is, however, a certain amount of statistical inference which can be accomplished within the CA framework,

Information-transforming versus statistical inference

as well as some interesting exploratory investigation of variability or stability of the maps, thanks to modern high-speed computing.

Stability of CA

By *stability* of the CA solution (the map, the inertias, the coordinates on specific principal axes, etc.), we are referring to the particular data set at hand, without reference to the population from which the data might come. Hence the issue of stability is relevant in all situations, even for population data or data obtained by convenience sampling. Here we assess how our interpretation is affected by the particular mix of row and column points determining the map. Would the map change dramatically (and thus our interpretation too) if one of the points is omitted? (for example, one of the species in our marine biology example, or one of the authors in the set of texts — see Chapter 10). This aspect of solution stability has already arisen several times when we discussed the concept of influence and how much each point influences the determination of the principal axes. In Chapter 11 the numerical *inertia contributions* were shown to provide indicators of the influence of each point. If a point contributes highly to an axis, then it is influential in the solution, which might be a problem if this point has a low mass, as discussed in Chapters 11 and 12. On the other hand, some points contribute very little to the solution, and could be removed without changing the map dramatically — that is, the map is *stable* with respect to including or removing these points. The acid test is to perform varous CAs, omitting selected points to see how the results are affected.

Sampling variability of the CA solution

Now looking outward beyond the data matrix, let us suppose that the data are collected by some sampling scheme from a wider *population*. For example, in the author data set of Exhibit 10.6 we know that the data represent a small part of the complete texts, and if the whole exercise were repeated on a different sample of each text, the counts of each letter would not be the same. It would be perfect, of course, if the sampling exercise could be repeated many times, and each time a CA performed to see if the features observed in the original map remained more or less the same or whether the books' and letters' positions changed. In other words, did what we see in the map arise by chance or was it a real feature of the 12 books being studied?

Bootstrapping the data

Since we cannot repeat the study, we have to rely on the actual data themselves to help us understand the sampling variability of the matrix. The usual way to proceed in statistics is to make assumptions about the population and then derive results about the uncertainty in the estimated values, which in our case are the coordinates of points in the map. A less formal way which avoids making any assumptions is provided by the *bootstrap**. The idea of the bootstrap is to regard the data as the population, since the data are the best

---

* The English expression "pulling yourself up by your own bootstraps" means using your own resources to get yourself out of a difficult situation.

estimate one has of the population. New data sets are created by resampling from the data in the same way as the data themselves were sampled. In the author data, the sampling has been performed for each text, not for each letter, so this is the way we should resample. For example, for the first book, "Three Daughters", 7144 letters were sampled, so we imagine — notionally, at least — these 7144 letters strung out in a long vector, in which there are 550 a s, 116 b s, 147 c 's, ... etc. Then we take a random sample of 7144 letters, *with replacement*, from this vector — the frequencies will not be exactly the same as those in the original table, but will reflect the variability that there is in those frequencies. This exercise is repeated for all the other rows of Exhibit 10.6, until we have a replicated table with the same row totals. This whole procedure can be repeated several times, usually between 100 and 1000 times, to establish many bootstrap replicates of the original data matrix.

An equivalent way to think about (and execute) the resampling is to make use of *multinomial sampling*. Each row profile defines a set of probabilities that can be regarded as the probability of obtaining an a, b, c, etc., in the respective text. Then it is a matter of sampling from a population with these probabilities, which is an easy computational algorithm, already implemented in R (see Computational Appendix). So we do not need to create the vector of 7144 letters for example, we need to use only the 26 probabilities of the letters in a multinomial sampling scheme.

Multinomial sampling

To illustrate the procedure on the author data, we first computed 100 replicates of the table by the sampling procedure described above. There are two ways to proceed now. The more difficult way is to repeat the CA on each replicate and then somehow compare the results to those obtained originally. The easier way is demonstrated here, called the *partial bootstrap*. Each replicated table can be regarded as a set of row profiles or set of column profiles, so the 100 replicated profiles are simply projected onto the CA map of the original data as supplementary points. Exhibit 25.1 shows the partial bootstrap of the 26 letters — each letter in larger font shows its original position in principal coordinates, with the 100 replicates in a tiny font. Usually we would not show all the replicates, but just show the *convex hull* of each set of points — this is the outer set of points connected by dotted lines in Exhibit 25.1, as if an elastic band has been placed around them.

Partial bootstrap of CA map, with convex hulls

Since the convex hull is sensitive to outlying replicates (for example, see the point for z on the right of Exhibit 25.1), it is usually *peeled*; that is, the convex hull of points is removed. The convex hull of the remaining points can be peeled again, and this process repeated until 5% of the outermost points in each subcloud have been removed. The convex hull of the remaining points is thus an estimate of a 95% confidence region for each letter. To make the estimation of these convex regions smoother, we generated 1000 replicates of each letter and then peeled off as close to 50 of them as possible (Exhibit 25.2),

Peeling the convex hull

Exhibit 25.1:
(Partial) bootstrap of 26 letters, after 100 replications of the data matrix. The more frequent the letter is in the texts, the more concentrated (less variable) are the replicates. Convex hulls are shown around each set of 100 replicated profiles.

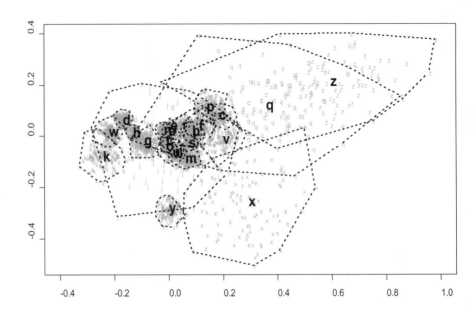

Exhibit 25.2:
Peeled convex hulls of points based on 1000 replicates (10 times more than in Exhibit 25.1), showing an approximate 95% confidence region for their distribution.

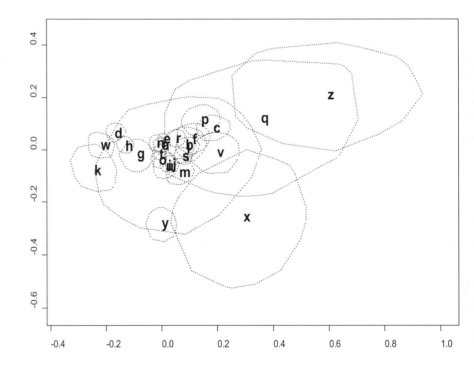

showing the convex hulls of the remaining subclouds. If two convex hulls do not overlap then this gives some assurance that the letters are significantly different in the texts. The actual level of significance is difficult to calculate because of the lack of formality in the procedure and the issue of multiple comparisons mentioned in Chapter 15. Fortunately, however, the procedure is conservative because of the projections onto the original map. If two convex hulls overlap in the map (for example, x and q), then it may still be possible that they do not overlap in the full space, but we would not be able to conclude this fact from the map. If they do not overlap in the projection (for example, k and y), then we know they do not overlap in the full space.

An alternative method for visualizing the confidence regions for each point in a CA map is to use confidence ellipses. These can be based on the replicates in the partial bootstrap above, or can be calculated making some theoretical assumptions. For example, the *delta method* uses the partial derivatives of the eigenvectors with respect to the multinomial proportions to calculate approximate variances and covariances of the coordinates. Then, assuming a bivariate normal distribution in the plane, confidence ellipses can be calculated — these enclose the true coordinates with 95% confidence, just like a confidence interval for single variables. This approach relies on the assumption of independent random sampling, which is not strictly satisfied in the author data because the occurrence of a particular letter is not independent of the occurrence of

The Delta method

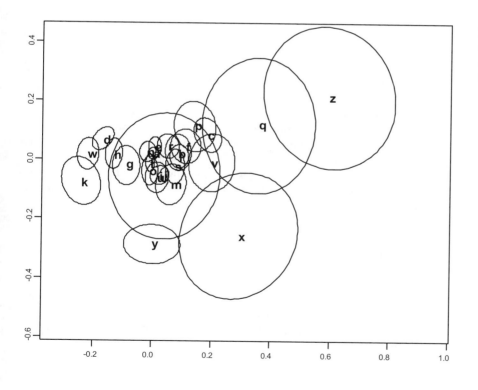

Exhibit 25.3:
Confidence ellipses based on the Delta method.

another (there is a similar problem in ecological sampling, where the same type of species seems to be found in groups in the samples). Nevertheless, the confidence ellipses for the letters in the author data are shown in Exhibit 25.3 and they bear a strong resemblance to the convex hulls in Exhibit 25.2, at least as far as overlapping is concerned.

Testing hypotheses — theoretical approach

The χ^2 test has been mentioned before as a test of independence on a contingency table. For example, the 5×3 table of Exhibit 4.1, which cross-tabulates 312 people on their level of readership and age group, has an inertia of 0.08326 and thus a χ^2 of $312 \times 0.08326 = 25.98$. Using the usual approximation to the χ^2 distribution, the P-value for the test is computed as 0.0035, a highly significant result. It is also possible to test the first principal inertia of a contingency table using other statistical approximations to the true distribution known as *asymptotic distributions*. The critical points for this test are exactly those that were used in Chapter 15 to test for significant clustering. The first principal inertia has value 0.07037, and its value as a χ^2 component is $312 \times 0.07037 = 21.96$. To test this value, refer to the table in the Theoretical Appendix, page 206, where the critical point (at a 5% level) is shown as 12.68 for a 5×3 table. Since 21.96 is much higher than this value we can conclude that the first dimension of the CA is significant and has not arisen by chance. The second principal inertia is more difficult to test, especially if we assume that the first principal inertia is significant, so we again resort to computer-based methods.

Testing hypotheses — Monte Carlo simulation

Given a hypothesis on the population, and knowing the way the data were sampled, we can set up a so-called *Monte Carlo simulation* to calculate the null distribution of the test statistic. For example, suppose we want to test both principal inertias of the readership data for significance. The null hypothesis is that there is no association between the rows and columns. The sampling here was not done as in the author data, where the text was sampled within each book — the analogy here would be that we sampled within each education group. By contrast, 312 people were sampled and then their education groups and readership categories were ascertained, so that the distribution of the education groups is also random, not fixed. Therefore we need to generate repeated samples of 312 people from the multinomial distribution which corresponds to the whole matrix, not row by row or column by column. The expected probabilities in each of the 15 cells of the table are equal to the product $r_i c_j$ of the masses. These define a vector of 15 probabilities under the null hypothesis which will be used to generate simulated multinomial samples of size 312. Two samples are given in Exhibit 25.4 alongside the original contingency table — in total 9999 tables were generated. For each simulated table the CA was performed and the two principal inertias calculated; hence along with the original observed value there are 10000 sets of values in total. Exhibit 25.5 shows the scatterplot of all of these, indicating the pair of values corresponding to the observed contingency table. It turns out there are only

EDUCATION GROUPS	Original data			1st Simulation			2nd Simulation			...
	C1	C2	C3	C1	C2	C3	C1	C2	C3	...
E1	5	7	2	2	9	5	4	5	7	...
E2	18	46	20	15	40	38	23	33	37	...
E3	19	29	39	13	36	27	17	34	25	...
E4	12	40	49	11	43	40	14	43	37	...
E5	3	7	16	8	12	13	5	12	16	...

Exhibit 25.4:
The original contingency table of Exhibit 4.1 and two of the 9999 simulated tables under the null hypothesis of no row–column association.

12 values out of 10000 that are larger than the observed first principal inertia; hence its P-value is estimated as 0.00012. For the second principal inertia there are 593 simulated values larger than the observed one, giving a P-value of 0.0593. At the 5% level the first is significant but not the second. At the

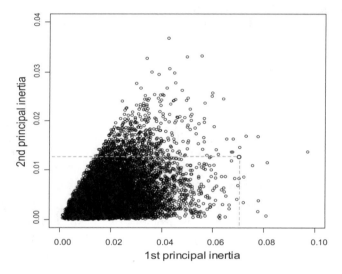

Exhibit 25.5:
Scatterplot of principal inertias from original CA and 9999 simulations of the 5×3 contingency table under the null hypothesis of no row–column association (two of the simulations are shown in Exhibit 25.4). The actual principal inertias are indicated by the larger circle and dashed lines..

same time we calculated the total inertia in each simulation — there are 19 simulated values larger than the observed total inertia of 0.08326. Therefore the P-value is 0.00019, which is our Monte Carlo estimate for the χ^2 test, compared to the P-value of 0.0035 that was based on the usual approximation to the χ^2 distribution.

Permutation tests (or randomization tests) are slightly different from the bootstrap and Monte Carlo procedures described above. For example, in the "blow-up" of the book points in Exhibit 10.7 we observed that the pairs of books by the same author lay in the same vicinity. It does seem unlikely that this could have occurred by chance, but what is the probability, or P-value, associated with this result? Thanks to modern computing this question can be answered as follows. First, calculate a measure of proximity between the pairs of books;

Permutation tests

an obvious measure is the sum of the six distances between pairs, which is equal to 0.4711. Then generate all possible ways of assigning the pairs of authors to the 12 texts; there are exactly $11 \times 9 \times 7 \times 5 \times 3 = 10395$ unique ways to rearrange them into six groups of 2. For each of these re-assignments of the labels to the points in the map, calculate the sum-of-distance measure. All these values define the *permutation distribution* of the test statistic, which has mean 0.8400 and standard deviation 0.1246 (in the Computational Appendix, page 253, the histogram of this distribution is shown). It turns out that there is no other assignment of the labels that gives a sum-of-distances smaller than the value observed in the CA map. Hence the P-value for the assertion that the pairs of texts are close is $P < 1/10395$, i.e., less than 0.0001, which is a highly significant result! Similar permutation tests were conducted for the subset CAs of the consonants and vowels separately (Exhibits 21.1 and 21.2), yielding $P = 0.0045$ and $P = 0.44$ respectively. Thus the consonants are accounting for the difference between the authors, not the vowels. Permutation tests are also routinely conducted in CCA to test whether the constrained space accounts for a significant part of the inertia. The idea is to compute a statistic such as the ratio of the constrained inertia to the unconstrained inertia and then perform a large number of CCAs, where in each one the rows of the explanatory variable matrix have been randomly permuted, generating a null distribution (see Computational Appendix, page 252–253).

SUMMARY:
*Aspects of
Stability and
Inference*

1. *Stability* concerns the data at hand and how much each row or column of data has influenced the display. The level of internal stability can be judged (a) by studying the row and column contributions and (b) by embarking on various re-analyses of the data which involve omitting single points or groups of points and seeing how the map is affected.

2. When the data are regarded as a sample of a wider population, the sampling variability can be investigated through a *bootstrap* resampling procedure to create replicates of the data table. The resampling should respect the row or column margins if these were fixed by the original sampling design.

3. In the *partial bootstrap* the row and/or column profiles of the replicated matrices are projected onto the CA solution as supplementary points. The replicate points can be summarized by drawing convex hulls or confidence ellipses.

4. Various theoretical approaches also exist, which rely on distributional assumptions in the population, for example the delta method and asymptotic theory based on normal approximations of the multinomial distribution.

5. *Monte Carlo* methods and *permutations tests* can be used to test specific hypotheses, relying on generating data under the null hypothesis to simulate (or calculate exactly) the null distribution of chosen test statistics, from which P-values can be deduced.

Theory of Correspondence Analysis

CA is based on fairly straightforward, classical results in matrix theory. The central result is the singular value decomposition (SVD), which is the basis of many multivariate methods such as principal component analysis, canonical correlation analysis, all forms of linear biplots, discriminant analysis and metric multidimensional scaling. In this appendix the theory of CA is summarized, as well as the theory of related methods discussed in the book. Matrix–vector notation is preferred because it is more compact, but also because it is closer to the implementation of the method in the R computing language.

Contents

Let $\mathbf{N}$ denote the $I \times J$ data matrix, with positive row and column sums (almost always $\mathbf{N}$ consists of nonnegative numbers, but there are some exceptions such as the one described at the end of Chapter 23). For notational simplicity the matrix is first converted to the *correspondence matrix* $\mathbf{P}$ by dividing $\mathbf{N}$ by its grand total $n = \sum_i \sum_j n_{ij} = \mathbf{1}^\mathsf{T}\mathbf{N}\mathbf{1}$ (the notation $\mathbf{1}$ is used for a vector of ones of length that is appropriate to its use; hence the first $\mathbf{1}$ is $I \times 1$ and the second is $J \times 1$ to match the row and column lengths of $\mathbf{N}$).

Correspondence matrix and preliminary notation

Correspondence matrix:

$$\mathbf{P} = \frac{1}{n}\mathbf{N} \tag{A.1}$$

The following notation is used:

Row and column masses:

$$r_i = \sum_{j=1}^{J} p_{ij} \qquad c_j = \sum_{i=1}^{I} p_{ij} \tag{A.2}$$

$$\text{i.e.,} \quad \mathbf{r} = \mathbf{P1} \qquad \mathbf{c} = \mathbf{P}^\mathsf{T}\mathbf{1}$$

Diagonal matrices of row and column masses:

$$\mathbf{D}_r = \mathrm{diag}(\mathbf{r}) \quad \text{and} \quad \mathbf{D}_c = \mathrm{diag}(\mathbf{c}) \tag{A.3}$$

Note that all subsequent definitions and results are given in terms of these relative quantities $\mathbf{P} = \{p_{ij}\}$, $\mathbf{r} = \{r_i\}$ and $\mathbf{c} = \{c_j\}$, whose elements add up to 1 in each case. Multiply these by n to recover the elements of the original matrix $\mathbf{N}$: $np_{ij} = n_{ij}$, $nr_i = i$-th row sum of $\mathbf{N}$, $nc_j = j$-th column sum of $\mathbf{N}$.

Basic computational algorithm — The computational algorithm to obtain coordinates of the row and column profiles with respect to principal axes, using the singular value decomposition (SVD), is as follows:

CA Step 1 — Calculate the matrix $\mathbf{S}$ of standarized residuals:

$$\mathbf{S} = \mathbf{D}_r^{-\frac{1}{2}}(\mathbf{P} - \mathbf{r}\mathbf{c}^\mathsf{T})\mathbf{D}_c^{-\frac{1}{2}} \tag{A.4}$$

CA Step 2 — Calculate the SVD of $\mathbf{S}$:

$$\mathbf{S} = \mathbf{U}\mathbf{D}_\alpha\mathbf{V}^\mathsf{T} \quad \text{where} \quad \mathbf{U}^\mathsf{T}\mathbf{U} = \mathbf{V}^\mathsf{T}\mathbf{V} = \mathbf{I} \tag{A.5}$$

where $\mathbf{D}_\alpha$ is the diagonal matrix of (positive) singular values in descending order: $\alpha_1 \geq \alpha_2 \geq \cdots$

CA Step 3 — Standard coordinates $\mathbf{\Phi}$ of rows:

$$\mathbf{\Phi} = \mathbf{D}_r^{-\frac{1}{2}}\mathbf{U} \tag{A.6}$$

CA Step 4 — Standard coordinates $\mathbf{\Gamma}$ of columns:

$$\mathbf{\Gamma} = \mathbf{D}_c^{-\frac{1}{2}}\mathbf{V} \tag{A.7}$$

CA Step 5 — Principal coordinates $\mathbf{F}$ of rows:

$$\mathbf{F} = \mathbf{D}_r^{-\frac{1}{2}}\mathbf{U}\mathbf{D}_\alpha = \mathbf{\Phi}\mathbf{D}_\alpha \tag{A.8}$$

CA Step 6 — Principal coordinates $\mathbf{G}$ of columns:

$$\mathbf{G} = \mathbf{D}_c^{-\frac{1}{2}}\mathbf{V}\mathbf{D}_\alpha = \mathbf{\Gamma}\mathbf{D}_\alpha \tag{A.9}$$

CA Step 7 — Principal inertias λ_k:

$$\lambda_k = \alpha_k^2, \quad k = 1, 2, \ldots, K \text{ where } K = \min\{I - 1, J - 1\} \tag{A.10}$$

The rows of the coordinate matrices in (A.6)–(A.9) refer to the rows or columns, as the case may be, of the original table, while the columns of these matrices refer to the principal axes, or dimensions, of which there are $\min\{I-1, J-1\}$, i.e., one less than the number of rows or columns, whichever is smaller. Notice how the principal and standard coordinates are scaled:

$$\mathbf{F}\mathbf{D}_r\mathbf{F}^\mathsf{T} = \mathbf{G}\mathbf{D}_c\mathbf{G}^\mathsf{T} = \mathbf{D}_\lambda \tag{A.11}$$

$$\mathbf{\Phi}\mathbf{D}_r\mathbf{\Phi}^\mathsf{T} = \mathbf{\Gamma}\,\mathbf{D}_c\,\mathbf{\Gamma}^\mathsf{T} = \mathbf{I} \tag{A.12}$$

That is, the weighted sum-of-squares of the principal coordinates on the k-th dimension (i.e., their inertia in the direction of this dimension) is equal to the principal inertia (or eigenvalue) $\lambda_k = \alpha_k^2$, the square of the k-th singular value, whereas the standard coordinates have weighted sum-of-squares equal to 1. All coordinate matrices have orthogonal columns, where the masses are always used in the calculation of the (weighted) scalar products.

The SVD is the fundamental mathematical result for CA, as it is for other dimension reduction techniques such as principal component analysis, canonical correlation analysis and linear discriminant analysis. This matrix decomposition expresses any rectangular matrix as a product of three matrices of simple structure, as in (A.5) above: $\mathbf{S} = \mathbf{U}\mathbf{D}_\alpha\mathbf{V}^\mathsf{T}$. The columns of the matrices $\mathbf{U}$ and $\mathbf{V}$ are the *left* and *right singular vectors* respectively, and the positive values α_k down the diagonal of $\mathbf{D}_\alpha$, in descending order, are the *singular values*. The SVD is related to the more well-known eigenvalue–eigenvector decomposition (or *eigendecomposition*) of a square symmetric matrix as follows: $\mathbf{S}\mathbf{S}^\mathsf{T}$ and $\mathbf{S}^\mathsf{T}\mathbf{S}$ are square symmetric matrices which have eigendecompositions $\mathbf{S}\mathbf{S}^\mathsf{T} = \mathbf{U}\mathbf{D}_\alpha^2\mathbf{U}^\mathsf{T}$ and $\mathbf{S}^\mathsf{T}\mathbf{S} = \mathbf{V}\mathbf{D}_\alpha^2\mathbf{V}^\mathsf{T}$, so the singular vectors are also eigenvectors of these respective matrices, and the singular values are the square roots of their eigenvalues. The practical utility of the SVD is that if one constructs another $I \times J$ matrix $\mathbf{S}_{(m)}$ from the the first m columns of $\mathbf{U}_{(m)}$ and $\mathbf{V}_{(m)}$ and the first m singular values in $\mathbf{D}_{\alpha(m)}$: $\mathbf{S}_{(m)} = \mathbf{U}_{(m)}\mathbf{D}_{\alpha(m)}\mathbf{V}_{(m)}^\mathsf{T}$, then $\mathbf{S}_{(m)}$ is the least-squares rank m approximation of $\mathbf{S}$ (this result is known as the *Eckart-Young theorem*). Since the objective of finding low-dimensional best-fitting subspaces coincides with the objective of finding low-rank matrix approximations by least-squares, the SVD solves our problem completely and in a very compact way. The only adaptation needed is to incorporate the weighting of the rows and columns by the masses into the SVD so that the approximations are by weighted least squares. If a generalized form of the SVD were defined, where the singular vectors are normalized with weighting by the masses, then the CA solution can be obtained in one step. For example, the generalized SVD of the *contingency ratios* $p_{ij}/(r_i c_j)$, elements of the matrix $\mathbf{D}_r^{-1}\mathbf{P}\mathbf{D}_c^{-1}$, centred at the constant value 1, leads to the standard row and column coordinates directly:

$$\mathbf{D}_r^{-1}\mathbf{P}\mathbf{D}_c^{-1} - \mathbf{1}\mathbf{1}^\mathsf{T} = \mathbf{\Phi}\mathbf{D}_\alpha\mathbf{\Gamma}^\mathsf{T} \quad \text{where} \quad \mathbf{\Phi}^\mathsf{T}\mathbf{D}_r\mathbf{\Phi} = \mathbf{\Gamma}^\mathsf{T}\mathbf{D}_c\mathbf{\Gamma} = \mathbf{I} \tag{A.13}$$

The bilinear   From steps 1 to 4 of the basic algorithm, the data in $\mathbf{P}$ can be written as
CA model   follows (see also (13.4) on page 101 and (14.9) on page 109):

$$p_{ij} = r_i c_j \left(1 + \sum_{k=1}^{K} \sqrt{\lambda_k} \phi_{ik} \gamma_{jk}\right) \tag{A.14}$$

(also called the *reconstitution formula*). In matrix notation:

$$\mathbf{P} = \mathbf{D}_r(\mathbf{1}\mathbf{1}^\mathsf{T} + \mathbf{\Phi}\mathbf{D}_\lambda^{1/2}\mathbf{\Gamma}^\mathsf{T})\mathbf{D}_c \tag{A.15}$$

Because of the simple relations (A.8) and (A.9) between the principal and
standard coordinates, this bilinear model can be written in several alternative
ways — see also (14.10) and (14.11) on pages 109–110.

Transition   The left and right singular vectors are related linearly, for example by multi-
equations between plying the SVD on the right by $\mathbf{V}$: $\mathbf{S}\mathbf{V} = \mathbf{U}\mathbf{D}_\alpha$. Expressing such relations in
rows and columns   terms of the principal and standard coordinates gives the following variations
                     of the same theme, called transition equations (see (14.1) & (14.2) and (14.5)
                     & (14.6) for the equivalent scalar versions):

Principal as a function of standard (barycentric relationships):

$$\mathbf{F} = \mathbf{D}_r^{-1}\mathbf{P}\mathbf{\Gamma} \qquad \mathbf{G} = \mathbf{D}_c^{-1}\mathbf{P}^\mathsf{T}\mathbf{\Phi} \tag{A.16}$$

Principal as a function of principal:

$$\mathbf{F} = \mathbf{D}_r^{-1}\mathbf{P}\mathbf{G}\mathbf{D}_\lambda^{-1/2} \qquad \mathbf{G} = \mathbf{D}_c^{-1}\mathbf{P}^\mathsf{T}\mathbf{F}\mathbf{D}_\lambda^{-1/2} \tag{A.17}$$

The equations (A.16) are those that were mentioned as early as Chapter 3,
which express the profile points as weighted averages of the vertex points,
where the weights are the profile elements. These are the equations that govern the *asymmetric maps*. The equations (A.17) show that the two sets of
principal coordinates, which govern the *symmetric map*, are also related by a
barycentric (weighted average) relationship, but with scale factors (the inverse
square roots of the principal inertias) that are different on each axis.

Supplementary   The transition equations are used to situate supplementary points on the map.
points   For example, given a supplementary column point with values in $\mathbf{h}$ ($I \times 1$),
                  divide by its total $\mathbf{1}^\mathsf{T}\mathbf{h}$ to obtain the column profile $\tilde{\mathbf{h}} = (1/\mathbf{1}^\mathsf{T}\mathbf{h})\mathbf{h}$ and then
                  use the profile transposed as a row vector in the second equation of (A.16),
                  for example, to calculate the coordinates $\mathbf{g}$ of the supplementary column:

$$\mathbf{g} = \tilde{\mathbf{h}}^\mathsf{T}\mathbf{\Phi} \tag{A.18}$$

Total inertia   The total inertia of the data matrix is the sum of squares of the matrix $\mathbf{S}$ in
and χ^2-distances (A.4):

$$\text{inertia} = \text{trace}(\mathbf{S}\mathbf{S}^\mathsf{T}) = \sum_{i=1}^{I}\sum_{j=1}^{J}\frac{(p_{ij} - r_i c_j)^2}{r_i c_j} \tag{A.19}$$

The inertia is also the sum of squares of the singular values, i.e., the sum of the eigenvalues:

$$\text{inertia} = \sum_{k=1}^{K} \alpha_k^2 = \sum_{k=1}^{K} \lambda_k \qquad (A.20)$$

The χ^2-distances between row profiles and between column profiles are:

$$\chi^2\text{-distances between rows } i \text{ and } i' : \sum_{j=1}^{J} \left(\frac{p_{ij}}{r_i} - \frac{p_{i'j}}{r_{i'}} \right)^2 / c_j \qquad (A.21)$$

$$\chi^2\text{-distances between columns } j \text{ and } j' : \sum_{i=1}^{I} \left(\frac{p_{ij}}{c_j} - \frac{p_{ij'}}{c_{j'}} \right)^2 / r_i \qquad (A.22)$$

To write the full set of χ^2-distances in the form of a square symmetric matrix requires a bit more work. First, calculate the matrix $\mathbf{A}$ of "χ^2 scalar products" between row profiles, for example, as:

$$\chi^2 \text{ scalar products between rows}: \quad \mathbf{A} = \mathbf{D}_r^{-1} \mathbf{P} \mathbf{D}_c^{-1} \mathbf{P}^\mathsf{T} \mathbf{D}_r^{-1} \qquad (A.23)$$

Then define the vector $\mathbf{a}$ as the elements on the diagonal of this matrix (i.e., the scalar products of the row profiles with themselves):

$$\mathbf{a} = \text{diag}(\mathbf{A}) \qquad (A.24)$$

Then the $I \times I$ matrix of squared χ^2-distances is:

$$\text{squared } \chi^2\text{-distance matrix between rows}: \mathbf{a}\mathbf{1}^\mathsf{T} + \mathbf{1}\mathbf{a}^\mathsf{T} - 2\mathbf{A} \qquad (A.25)$$

To calculate the $J \times J$ matrix of squared χ^2-distances between column profiles, interchange rows with columns in (A.23), defining $\mathbf{A}$ as $\mathbf{D}_c^{-1} \mathbf{P}^\mathsf{T} \mathbf{D}_r^{-1} \mathbf{P} \mathbf{D}_c^{-1}$ and then following with (A.24) and (A.25).

The contributions of the row and columns points to the inertia on the k-th dimension are the inertia components: *Contributions of points to principal inertias*

$$\text{for row } i: \quad \frac{r_i f_{ik}^2}{\lambda_k} = r_i \phi_{ik}^2 \qquad \text{for column } j: \quad \frac{c_j g_{jk}^2}{\lambda_k} = c_j \gamma_{jk}^2 \qquad (A.26)$$

recalling the relationship between principal and standard coordinates given in (A.8) and (A.9): $f_{ik} = \sqrt{\lambda_k}\phi_{ik}$, $g_{jk} = \sqrt{\lambda_k}\gamma_{jk}$ (notice that the square roots of the values in (A.26) are exactly the coordinates proposed for the standard CA biplot of Chapter 13, which shows that the squared lengths of these coordinates are the contributions to the principal axes).

The contributions of the dimensions to the inertia of the i-th row and j-th column points (i.e., the squared cosines or squared correlations) are: *Contributions of principal axes to point inertias (squared correlations)*

$$\text{for row } i: \quad \frac{f_{ik}^2}{\sum_k f_{ik}^2} \qquad \text{for column } j: \quad \frac{g_{jk}^2}{\sum_k g_{jk}^2} \qquad (A.27)$$

As shown in Chapter 11, the denominators in (A.27) are the squared χ^2-distances between the corresponding profile point and the average profile.

Ward
clustering of row
or column profiles The clustering of Chapter 15 is described here in terms of the rows; exactly the same applies to the clustering of the columns. The rows are clustered at each step of the algorithm to minimise the decrease in the χ^2 statistic (equivalently, the decrease in the inertia since inertia $= \chi^2/n$, where n is the total of the table). This clustering criterion is equivalent to Ward clustering, where each cluster is weighted by the total mass of its members. The measure of difference between rows can be shown to be the weighted form of the squared chi-squared distance between profiles. Suppose $\mathbf{a}_i$ and r_i, $i = 1, \ldots, I$, denote the I row profiles of the data matrix, and their masses, respectively. Then identifying the pair that gives the least decrease in inertia is equivalent to looking for the pair of rows (i, i') which minimize the following measure:

$$\frac{r_i r_{i'}}{r_i + r_{i'}} \|\mathbf{a}_i - \mathbf{a}_{i'}\|_c^2 \tag{A.28}$$

The two rows are then merged by summing their frequencies, and the profile and mass are recalculated. The same measure of difference as (A.28) is calculated at each stage of the clustering for the row profiles at that stage (see (15.2) on page 120 for the equivalent formula based on profiles of clusters), and the two profiles with the least difference are merged. So (A.28) is the level of clustering in terms of the inertia decrease, or if multiplied by n the decrease in χ^2. In the case of a contingency table the level of clustering can be tested for significance using the tables at the end of this Appendix.

Stacked tables Suppose tables $\mathbf{N}_{qs}$, $q = 1, \ldots, Q$, $s = 1, \ldots, S$ are concatenated row- and/or columnwise to make a block matrix $\mathbf{N}$. If the marginal frequencies are the same in each row and in each column (as is the case when the same individuals are cross-tabulated separately in several tables), then the inertia of $\mathbf{N}$ is the average of the separate inertias of the tables $\mathbf{N}_{qs}$:

$$\text{inertia}(\mathbf{N}) = \frac{1}{QS} \sum_{q=1}^{Q} \sum_{s=1}^{S} \text{inertia}(\mathbf{N}_{qs}) \tag{A.29}$$

Multiple CA Suppose the original matrix of categorical data is $N \times Q$, i.e., N cases and Q variables. Classical multiple CA (MCA) has two forms. The first form converts the cases-by-variables data to an indicator matrix $\mathbf{Z}$ where the categorical data have been recoded as dummy variables. If the q-th variable has J_q categories, this indicator matrix will have $J = \sum_q J_q$ columns (see Chapter 18, Exhibit 18.1 for an example). Then the indicator version of MCA is the application of the basic CA algorithm to the matrix $\mathbf{Z}$, resulting in coordinates for the N cases and the J categories. The second form of MCA calculates the Burt matrix $\mathbf{B} = \mathbf{Z}^\mathsf{T}\mathbf{Z}$ of all two-way cross-tabulations of the Q variables (see Chapter 18, Exhibit 18.4 for an example). Then the Burt version of MCA is the application of the basic CA algorithm to the matrix $\mathbf{B}$, resulting in coordinates for the J categories ($\mathbf{B}$ is a symmetric matrix). The standard coordinates of the categories are identical in the two versions of MCA, and the principal inertias in the Burt version are the squares of those in the indicator version.

Joint CA (JCA) is the fitting of the off-diagonal cross-tabulations of the Burt matrix, ignoring the cross-tabulations on the block diagonal. The algorithm we use is an alternating least-squares procedure which successively applies CA to the Burt matrix which has been modified by replacing the values on the block diagonal with estimated values from the CA of the previous iteration. The algorithm itself is explained in more detail in the Computational Appendix. On convergence of the JCA algorithm, the CA is performed on the last modified Burt matrix, $\widetilde{\mathbf{B}}$, which has its diagonal blocks perfectly fitted by construction. In other words, supposing that the solution requested is two-dimensional, then the modified diagonal blocks satisfy (A.14) exactly using just two terms in the bilinear CA model (or reconstitution formula).

Joint CA

Hence the total inertia of $\widetilde{\mathbf{B}}$ includes a part Δ for these diagonal blocks, and so do the first two principal inertias, $\tilde{\lambda}_1$ and $\tilde{\lambda}_2$, which perfectly explain the part Δ. To obtain the percentage of inertia explained by the two-dimensional solution, the amount Δ has to be discounted both from the total and from the sum of the two principal inertias. The value of Δ can be obtained via the difference between the inertia of the original Burt matrix $\mathbf{B}$ (whose diagonal inertias are known) and the modified one $\widetilde{\mathbf{B}}$, as follows (here we use the result (A.29) which applies to the subtables of $\mathbf{B}$, denoted by $\mathbf{B}_{qs}$, and those of $\widetilde{\mathbf{B}}$), whose off-diagonal tables are the same):

Percentage of inertia explained in JCA

$$
\begin{aligned}
\text{inertia}(\mathbf{B}) &= \frac{1}{Q^2}\left(\sum\sum_{q\neq s}\text{inertia}(\mathbf{B}_{qs}) + \sum_q \text{inertia}(\mathbf{B}_{qq})\right) \\
&= \frac{1}{Q^2}\left(\sum\sum_{q\neq s}\text{inertia}(\mathbf{B}_{qs}) + (J - Q)\right) \\
\text{inertia}(\widetilde{\mathbf{B}}) &= \frac{1}{Q^2}\left(\sum\sum_{q\neq s}\text{inertia}(\mathbf{B}_{qs})\right) + \Delta
\end{aligned}
$$

Subtracting the above leads to:

$$
\text{inertia}(\mathbf{B}) - \text{inertia}(\widetilde{\mathbf{B}}) = \frac{J - Q}{Q^2} - \Delta \tag{A.30}
$$

which gives the value of Δ:

$$
\Delta = \frac{J - Q}{Q^2} - \left(\text{inertia}(\mathbf{B}) - \text{inertia}(\widetilde{\mathbf{B}})\right) \tag{A.31}
$$

Discounting this amount from the total and the sum of the principal inertias (assuming a two-dimensional solution) gives the percentage of inertia explained by the JCA solution:

$$
100 \times \frac{\tilde{\lambda}_1 + \tilde{\lambda}_2 - \Delta}{\text{inertia}(\widetilde{\mathbf{B}}) - \Delta} \tag{A.32}
$$

The previous section showed how to discount the extra inertia as a result of the modified diagonal blocks of the Burt matrix in JCA. There is an identical situation at the level of each point. Each category point j has an additional amount of inertia, δ_j, due to the modified diagonal blocks. In the case of the original Burt matrix $\mathbf{B}$ we know exactly what this extra amount is due to the diagonal matrices in the diagonal blocks: for the j-th point it is $(1 - Qc_j)/Q^2$, where c_j is the j-th mass (summing these values for $j = 1, \ldots, J$, we obtain $(J - Q)/Q^2$ which was the total additional amount due to the diagonal blocks of $\mathbf{B}$). Therefore, just as above, we can derive how to obtain contributions of the two-dimensional solution to the point inertias as follows:

$$\text{inertia}(j\text{-th category of } \mathbf{B}) = \text{off-diagonal components} + \frac{1 - Qc_j}{Q^2}$$

$$\text{inertia}(j\text{-th category of } \widetilde{\mathbf{B}}) = \text{off-diagonal components} + \delta_j$$

Subtracting the above (the "off-diagonal components" are the same) leads to:

$$\text{inertia}(j\text{th category of } \mathbf{B}) - \text{inertia}(j\text{-th category of } \widetilde{\mathbf{B}}) = \frac{1 - Qc_j}{Q^2} - \delta_j$$

which gives the value of δ_j:

$$\delta_j = \frac{1 - Qc_j}{Q^2} - \left(\text{inertia}(j\text{-th category of } \mathbf{B}) - \text{inertia}(j\text{-th category of } \widetilde{\mathbf{B}}) \right)$$
(A.33)

Discounting this amount from the j-th category's inertia and similarly from the sum of the components of inertia in two dimensions gives the relative contributions (qualities) with respect to the two-dimensional JCA solution:

$$\frac{c_j \tilde{g}_{j1}^2 + c_j \tilde{g}_{j2}^2 - \delta_j}{(\sum_k c_j \tilde{g}_{jk}^2) - \delta_j}$$
(A.34)

where $\tilde{g}_{jk}$ is the principal coordinate of category j on axis k in the CA of $\widetilde{\mathbf{B}}$ (JCA solution), and the summation in the denominator is for all the dimensions. Notice that $\sum_j \delta_j = \Delta$ (i.e., summing (A.33) gives (A.31)).

The MCA solution can be adjusted to optimize the fit to the off-diagonal tables (this could be called a JCA *conditional on* the MCA solution). The optimal adjustments can be determined by weighted least-squares, as described in Chapter 19, but the problem is that the solution is not nested. So we prefer slightly sub-optimal adjustments which retain the nesting property and are very easy to compute from the MCA solution of the Burt matrix. The adjustments are made as follows (see Chapter 19, pages 140–141, for an illustration):

Adjusted total inertia of Burt matrix:

$$\text{adjusted total inertia} = \frac{Q}{Q - 1} \times \left(\text{inertia of } \mathbf{B} - \frac{J - Q}{Q^2} \right)$$
(A.35)

Adjusted principal inertias (eigenvalues) of Burt matrix:

$$\lambda_k^{\text{adj}} = \left(\frac{Q}{Q-1}\right)^2 \times \left(\sqrt{\lambda_k} - \frac{1}{Q}\right)^2, \quad k = 1, 2, \ldots \qquad (A.36)$$

Here λ_k refers to the k-th principal inertia of the Burt matrix; hence $\sqrt{\lambda_k}$ is the k-th principal inertia of the indicator matrix. The adjustments are made only to those dimensions for which $\sqrt{\lambda_k} > \frac{1}{Q}$ and no further dimensions are used — hence percentages of inertia do not add up to 100%. It can be proved that these percentages are lower bound estimates of those that are obtained in a JCA, and in practice they are close to the JCA percentages.

Subset CA is simply the application of the same CA algorithm to a selected part of the standardized residual matrix $\mathbf{S}$ in (A.4) (*not* to the subset of the original matrix). The masses of the full matrix are thus retained and all subsequent calculations are the same, except they are applied to the subset. Suppose that the columns are subsetted, but not the rows. Then the rows still maintain the centring property of CA; i.e., their weighted averages are at the origin of the map, whereas the columns are no longer centred. Subset MCA is performed by applying subset CA on a submatrix of the indicator matrix or the Burt matrix. In the case of the Burt matrix, a selection of categories implies that this subset has to be specified for both the rows and columns.

Subset CA and subset MCA

If the data matrix $\mathbf{N}$ is square asymmetric, where both rows and columns refer to the same objects, then $\mathbf{N}$ can be written as the sum of symmetric and skew-symmetric parts:

Analysis of square asymmetric tables

$$\mathbf{N} = \frac{1}{2}(\mathbf{N} + \mathbf{N}^{\mathsf{T}}) + \frac{1}{2}(\mathbf{N} - \mathbf{N}^{\mathsf{T}}) \qquad (A.37)$$
$$= \text{symmetric} + \text{skew-symmetric}$$

CA is applied to each part separately, but with a slight variation for the skew-symmetric part. The analysis of the symmetric part $\frac{1}{2}(\mathbf{N} + \mathbf{N}^{\mathsf{T}})$ is the usual CA — this provides one set of coordinates, and the masses are the averages of the row and column masses corresponding to the same object: $w_i = \frac{1}{2}(r_i + c_i)$. The analysis of the skew-symmetric part $\frac{1}{2}(\mathbf{N} - \mathbf{N}^{\mathsf{T}})$ is the application of the CA algorithm *without centring* and using the same masses as in the symmetric analysis; i.e., the "standardized residuals" matrix of (A.4) is rather the "standardized differences" matrix

$$\mathbf{S} = \mathbf{D}_w^{-\frac{1}{2}}[\frac{1}{2}(\mathbf{P} - \mathbf{P}^{\mathsf{T}})]\mathbf{D}_w^{-\frac{1}{2}} \qquad (A.38)$$

where $\mathbf{P}$ is the correspondence matrix and $\mathbf{D}_w$ is the diagonal matrix of the masses w_i. As described in Chapter 22, both these analyses are subsumed in the ordinary CA of the block matrix

$$\begin{bmatrix} \mathbf{N} & \mathbf{N}^{\mathsf{T}} \\ \mathbf{N}^{\mathsf{T}} & \mathbf{N} \end{bmatrix} \qquad (A.39)$$

If **N** is an $I \times I$ matrix, then the $2I - 1$ dimensions which emanate from this CA can be easily allocated to the symmetric and skew-symmetric solutions since the symmetric dimensions have unique principal inertias while the skew-symmetric dimensions occur in pairs with equal principal inertias. Similarly the coordinate vectors for each dimension have two parts: for dimensions corresponding to the symmetric analysis these are simple repeats on each other, while for dimensions corresponding to the skew-symmetric analysis these are repeats with a change of sign (see Chapter 22 for an example).

Canonical CA   In canonical CA (CCA) an additional matrix **X** of explanatory (independent) variables is available, and the requirement is that the dimensions be linearly related to **X**. The total inertia is split into two parts: a part that is linearly related to the independent variables, called the inertia in the *constrained space*, and a part that is not, the inertia in the *unconstrained space*. CCA is necessarily an "asymmetric" method since **X** is an additional set of either rows or columns. The usual data structure is that the rows are sampling units and **X** is an additional set of M columns, i.e., $I \times M$. There is a regression step in CCA which calculates the $I \times J$ constrained matrix, whose columns are linearly related to **X**. The difference between **P** and the constrained matrix is the unconstrained matrix, whose columns are not linearly ~~un~~related to **X**. CCA thus consists of applying CA to the constrained matrix and (optionally) to the unconstrained matrix. In each application the original row and column masses are maintained for all computations, and the various results such as coordinates, principal inertias, contributions, reconstruction formula, etc., are the same as in a regular application of CA. We assume that the columns of **X** are standardized, using the row masses as weights in the calculation of means and variances. If there are some categorical independent variables, these are coded as dummy variables, dropping one category of each variable as in a conventional regression analysis. The retained dummy variables are then also standardized in the ~~usual way~~. same way as the columns of X

The steps in CCA are as follows:

CCA Step 1 — Calculate the standardized residuals matrix **S** *as in CA:*

$$\mathbf{S} = \mathbf{D}_r^{-\frac{1}{2}}(\mathbf{P} - \mathbf{rc}^\mathsf{T})\mathbf{D}_c^{-\frac{1}{2}} \tag{A.40}$$

CCA Step 2 — Calculate the $I \times I$ projection matrix, of rank M, which projects onto the constrained space:

$$\mathbf{Q} = \mathbf{D}_r^{\frac{1}{2}}\mathbf{X}(\mathbf{X}^\mathsf{T}\mathbf{D}_r\mathbf{X})^{-1}\mathbf{X}^\mathsf{T}\mathbf{D}_r^{\frac{1}{2}} \tag{A.41}$$

CCA Step 3 — Project the standardized residuals to obtain the constrained matrix:

$$\mathbf{S}^\star = \mathbf{QS} \tag{A.42}$$

CCA Step 4 — Apply CA Steps 1–6 (page 202) to **S***:

CCA Step 5 — Principal inertias $\lambda_k^\star$ in constrained space:

$$\lambda_k^\star = \alpha_k^2, \quad k = 1, 2, \ldots, K \text{ where } K = \min\{I - 1, J - 1, M\} \tag{A.43}$$

CCA Step 5 (optional) — Project the standardized residuals onto the unconstrained space:

$$\mathbf{S}^{\perp} = (\mathbf{I} - \mathbf{Q})\mathbf{S} = \mathbf{S} - \mathbf{S}^{\star} \tag{A.44}$$

CCA Step 6 (optional) — Apply CA Steps 1–6 to $\mathbf{S}^{\perp}$.

As described in Chapter 24, the principal inertias in (A.43) can be expressed as percentages of the total inertia, or as percentages of the constrained inertia, which is the sum of squares of the elements in $\mathbf{S}^{\star}$, equal to $\sum_k \lambda_k^{\star}$.

Tables for testing for significant clustering or significant dimensions

In the case of a contingency table based on a random sample, the first principal inertia can be tested for statistical significance. This is the same test as was used in the case of the Ward clustering of Chapter 15. In that case a critical level for clustering, on the χ^2 scale, can be determined from the tables in Exhibit A.1 below, according to the size of the table (see page 119 for the food store example, a 5×4 table for which the critical point in Exhibit A.1 is 15.24). These critical points are the same for testing the first principal inertia for significance. For example, in the same example of the food stores, given in Exhibit 15.3, the first principal inertia was 0.02635, which if expressed as a χ^2 component is $0.02635 \times 700 = 18.45$. Since 18.45 is greater than the critical point 15.24, the first principal inertia is statistically significant (at the 5% level).

Exhibit A.1: *Critical values for multiple comparisons test on a $I \times J$ (or $J \times I$) contingency table. The same critical points apply to testing the significance of a principal inertia. Significance level is 5%.*

					J				
I	3	4	5	6	7	8	9	10	11
3	8.59								
4	10.74	13.11							
5	12.68	15.24	17.52						
6	14.49	17.21	19.63	21.85					
7	16.21	19.09	21.62	23.95	26.14				
8	17.88	20.88	23.53	25.96	28.23	30.40			
9	19.49	22.62	25.37	27.88	30.24	32.48	34.63		
10	21.06	24.31	27.15	29.75	32.18	34.50	36.70	38.84	
11	22.61	25.96	28.90	31.57	34.08	36.45	38.72	40.91	43.04
12	24.12	27.58	30.60	33.35	35.93	38.36	40.69	42.93	45.10
13	25.61	29.17	32.27	35.09	37.73	40.22	42.60	44.90	47.12

Source: Pearson, E.S. & Hartley, H.O. (1972). *Biometrika Tables for Statisticians, Volume 2: Table 51.* Cambridge University Press, UK.

Computation of Correspondence Analysis

In this appendix the computation of CA is illustrated using the object-oriented computing language R, which can be freely downloaded from the website:

http://www.r-project.org

We assume here that the reader has some basic knowledge of this language, which has become the *de facto* standard for statistical computing. If not, the above website gives many resources for learning it. The scripts which are given in this appendix are available at the website of the CARME network:

http://www.carme-n.org

(CARME = Correspondence Analysis and Related MEthods). At the end of this appendix we shall also comment on commercially available software, and describe different graphical options for producing maps.

Contents

The R program

The R program provides all the tools necessary to produce CA maps, the most important one being the singular value decomposition (SVD). These tools are encapsulated in R *functions*, and several functions and related material can be gathered together to form an R *package*. An R package called **ca** is already available for doing every type of CA described in this book, to be demonstrated later in this appendix. But before that, we show step-by-step how to perform various computations using R. The three-dimensional graphics package **rgl** will also be demonstrated in the process. In the following we use a Courier font for all R instructions and R output; for example, here we create the matrix (13.2) on page 99, calculate its SVD and store it in an R "svd" object, and then ask for the part of the object labelled 'd' (the singular values):

```
> table.T   <- matrix(c(8,5,-2,2,4,2,0,-3,3,6,2,3,3,-3,-6,
+               -6,-4,1,-1,-2),nrow=5)
> table.SVD <- svd(table.T)
> table.SVD$d
[1] 1.412505e+01 9.822577e+00 6.351831e-16 3.592426e-33
```

The commands are indicated in slanted script (the prompt > is not typed), while the results are given in regular typewriter script. A + at the start of the line indicates continuation of the command.

213

Entering data into R

Entering data into R has its peculiarities, but once you have managed to do it, the rest is easy! The `read.table()` function is one of the most useful ways to input data matrices, and the easiest data sources are a text file or an Excel file. For example, suppose we want to input the 5×3 data table on readership given in Exhibit 3.1. Here are three options for reading it in.

1. Suppose the data are in a text file as follows:

   ```
         C1    C2    C3
   E1     5     7     2
   E2    18    46    20
   E3    19    29    39
   E4    12    40    49
   E5     3     7    16
   ```

   and suppose the file is called `reader.txt` and stored in the present R working directory. Then execute the following statement in R:

   ```
   > read.table("reader.txt")
   ```

2. An easier alternative is to copy the above file into the clipboard by selecting the contents in the text- or word-processor and copying using the pull-down Edit menu or right-clicking the mouse (assuming Windows platform). Then execute a similar command by reading directly from the clipboard:

   ```
   > read.table("clipboard")
   ```

3. In a similar fashion, data can be read from an Excel file* via the clipboard, assuming the data are in an Excel file, as displayed below:

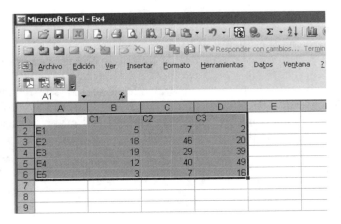

   The cells of this table have been selected and then copied. The command

   ```
   > table <- read.table("clipboard")
   ```

   results in the table being stored as an R "data frame" object with the name

---

* Using the R package **foreign**, distributed with the program, it is possible to read other data formats, e.g., Stata, Minitab, SPSS, SAS, Systat and DBF.

table. Notice that the success of this `read.table()` command relies on the fact that the first line of the copied table contains one less entity than the other lines — this is why there is an empty cell in the top left-hand corner of the Excel table, similarly in the text file. If the `read.table()` function finds one less entity in the first line, it realizes that the first line consists of column labels and the subsequent lines have the column labels in the first column. The contents of `table` can be seen by entering

```
> table
   C1 C2 C3
E1  5  7  2
E2 18 46 20
E3 19 29 39
E4 12 40 49
E5  3  7 16
```

The object includes the row and column names, which can be accessed by typing `rownames(table)` and `colnames(table)`, for example:

```
> rownames(table)
[1] "E1" "E2" "E3" "E4" "E5"
```

We now describe systematically the computations for each chapter, starting with Chapter 2. Only basic R functions and the three-dimensional plotting package **rgl**[†] will be used, leaving till later a demonstration of the **ca** package which does the calculations in a much more compact way.

R scripts for each chapter

In Chapter 2 we showed some triangular plots of the travel data set. Suppose that the profile data of Exhibit 2.1 are input as described before and stored in the data frame `profiles` in R; that is, after copying the profile data:

Chapter 2: Profiles and the Profile Space

```
> profiles <- read.table("clipboard")
> profiles
              Holidays HalfDays FullDays
Norway           0.333    0.056    0.611
Canada           0.067    0.200    0.733
Greece           0.138    0.862    0.000
France/Germany   0.083    0.083    0.833
```

(notice that the column names had been originally written without blanks, otherwise the data would not have been read correctly). We can do a three-dimensional view of the profiles using the **rgl** package as follows (assuming **rgl** has been installed and loaded — see the footnote below).

*Example of three-dimensional graphics using **rgl** package*

```
> rgl.lines(c(0,1.2), c(0,0), c(0,0))
> rgl.lines(c(0,0), c(0,1.2), c(0,0))
> rgl.lines(c(0,0), c(0,0), c(0,1.2))
> rgl.lines(c(0,0), c(0,1), c(1,0), size=2)
```

[†] The **rgl** package is not one of the packages provided as standard with R, but needs to be installed by downloading it from the R website or `www.carme-n.org`.

Exhibit B.1:
*Three-dimensional
view of the country
row profiles of the
travel data set,
using the R package
rgl.*

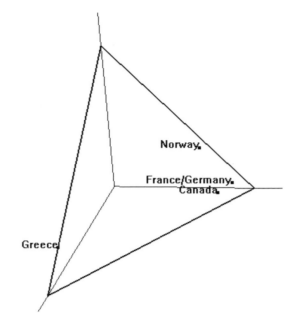

Exhibit B.2:
*Rotation of the
three-dimensional
space to show the
triangle in which the
profile points lie.*

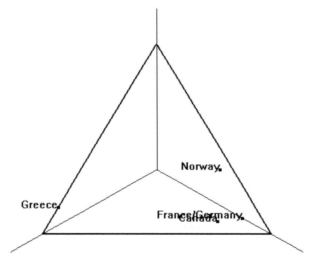

```
> rgl.lines(c(0,1), c(1,0), c(0,0), size=2)
> rgl.lines(c(0,1), c(0,0), c(1,0), size=2)
> rgl.points(profiles[,3],profiles[,1],profiles[,2],
+            size=4)
> rgl.texts(profiles[,3],profiles[,1],profiles[,2],
+            text=row.names(profiles))
```

The 3-D scatterplot from a certain viewpoint is shown in Figure B.1. Using the mouse while pressing the left button, this figure can be rotated to give a realistic three-dimensional feeling. Figure B.2 shows one of these rotations where the viewpoint is flat onto the triangle that contains the profile points. The mouse wheel allows zooming into the display.

As an illustration of the graphics in Chapter 3, we give the code to draw the triangular coordinate plot in Exhibit 3.2 using R, which needs some trignometry to calculate the (x, y) positions of each point. Assuming the table has been read as on the bottom of page 214 into the data frame `table`, the following commands in R produce the figure in Exhibit B.3. The first statement calculates the row profiles in `table.pro` using the `apply()` function — this function can operate on rows or columns; here the parameter "1" indicates rows, and "sum" the operation required. R stores matrices as a string of columns, so the division of `table` by the row sums does just the right thing, dividing the first column by the row sums, then the second and so on. The two subsequent statements calculate the x and y coordinates of the five profiles in an equilateral triangle of side 1, using the first and third profile values (only two out of the three are needed to situate the point).

Chapter 3: Masses and Centroid

`apply()` *function*

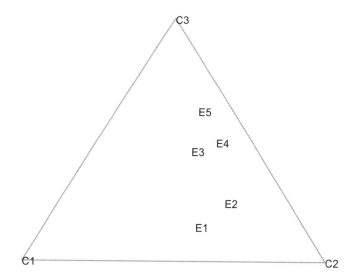

Exhibit B.3:
Plot of five education group profiles in the triangular coordinate space.

```
> table.pro <- table / apply(table, 1, sum)
> table.x <- 1 - table.pro[,1] - table.pro[,3] / 2
> table.y <- table.pro[,3] * sqrt(3) / 2
> plot.new()
> lines(c(0,1,0.5,0), c(0,0,sqrt(3)/2,0), col="gray")
> text(c(0,1,0.5), c(0,0,sqrt(3)/2), labels=colnames(table))
> text(table.x, table.y, labels=rownames(table))
```

Example of two-dimensional graphics

Chapter 4:
Chi-square
Distances and
Inertia

In Chapter 4 the χ^2 statistic, inertia and χ^2-distances were calculated, each of which is illustrated here. The calculations are performed on the readership data frame `table` used previously.

— χ^2 *statistic and total inertia:*

```
> table.rowsum <- apply(table, 1, sum)
> table.colsum <- apply(table, 2, sum)
> table.sum <- sum(table)
> table.exp <- table.rowsum %o% table.colsum / table.sum
> chi2 <- sum((table - table.exp)^2 / table.exp)
> chi2
[1] 25.97724
> chi2 / table.sum
[1] 0.08326039
```

Outer product
operator %o%

Notice the use of the *outer product* operator `%o%` in the fourth command above; this multiplies every element in the vector to the left of the operator with every element in the vector on the right.

— χ^2-*distances of row profiles to centroid:*

We first show the least elegant (but most obvious) way of calculating the square of the χ^2-distance for the fifth row of the table, as shown in brackets in (4.4). A `for` loop in R is used to build up the sum of the three terms:

```
> chidist <- 0
```

Example of `for`
loop in R

```
> for(j in 1:3) {
+   chidist<-chidist+
+   (table.pro[5,j] - table.colmass[j])^2 / table.colmass[j]
+ }
> chidist
        C1
0.1859165
```

The label `C1` is given to the value of `chidist` probably because this is the first column of the loop. A more elegant way is to compute all five distances in one step. We need to subtract the row of column masses from each row of the profile matrix, square these differences, and then divide each row again by the column masses, finally adding up the rows. Row operations are slightly more difficult in R because matrices are stored as column vectors, so one solution is to transpose the profile matrix first, using the `t()` transpose function. Then all the columns of the transposed object (previously rows) are summed, using the `apply()` function with parameters `2,sum` indicating column sums:

Tranpose function
`t()`

```
> apply((t(table.pro)-table.colmass)^2 / table.colmass,2,sum)
        E1         E2         E3         E4         E5
0.35335967 0.11702343 0.02739229 0.03943842 0.18591649
```

Distance function
`dist()`

Finally, all χ^2-distances can be computed, between all profiles and in particular between the profiles and their average, using the `dist()` function, which by default computes a Euclidean distance matrix between rows of a matrix.

First, the row of column masses (average row profile) is appended to the profile matrix using the function `rbind()` (row binding) to form a 6×3 profile matrix `tablec.pro`. Second, we need to divide each profile element by the corresponding square root of the average. An alternative to having to use the transpose operation again is to use the versatile `sweep()` function, which acts like `apply()` but has more options. In the second command below the options of `sweep` are 2 (operating down the columns), `sqrt(table.colmass)` (the vector used for the operation) and `"/"` (the operation is division):

Row binding using `rbind()`

Versatile `sweep()` *function*

```
> tablec.pro <- rbind(table.pro, table.colmass)
> rownames(tablec.pro)[6]<-"ave"
> dist(sweep(tablec.pro, 2, sqrt(table.colmass), FUN="/"))
          E1         E2         E3         E4         E5
E2   0.3737004
E3   0.6352512 0.4696153
E4   0.7919425 0.5065568 0.2591401
E5   1.0008054 0.7703644 0.3703568 0.2845283
ave  0.5944406 0.3420869 0.1655062 0.1985911 0.4311803
```

The last line, which was labelled "ave" for the appended average profile (see second command above), gives the χ^2-distances (square roots of the squared values calculated above). All other distances between the five row profiles are also given, and the result of `dist()` is an R distance object which stores only the triangular matrix of distances.

In Chapter 5 the χ^2-distances are visualized by stretching the coordinate axes by amounts inversely proportional to the square roots of the corresponding masses. So this is a similar sequence of code to that given previously for the three-dimensional plot of Chapter 2, except that each coordinate is divided by `sqrt(table.colmass)`. The trickier aspect is to decide which profile elements go with which dimensions, to reproduce Exhibit 5.2 — we leave this as a small exercize for the reader, but the actual script is given on the website.

Chapter 5:
Plotting
Chi-square
Distances

Chapter 6 starts with actual CAs in that dimension-reduction is involved, so here we shall perform our first singular value decomposition (SVD). We first input the health self-assessment data set and call it `health`; then we follow the steps given on page 202 of the Theoretical Appendix. The preparatory steps (A.1–3) are as follows:

Chapter 6:
Reduction of
Dimensionality

```
> health.P <- health / sum(health)
> health.r <- apply(health.P, 1, sum)
> health.c <- apply(health.P, 2, sum)
> health.Dr <- diag(health.r)
> health.Dc<-diag(health.c)
> health.Drmh <- diag(1/sqrt(health.r))
> health.Dcmh <- diag(1/sqrt(health.c))
```

Diagonal matrix
function `diag()`

The last two commands above create $\mathbf{D}_r^{-\frac{1}{2}}$ and $\mathbf{D}_c^{-\frac{1}{2}}$, respectively, since we need them repeatedly later (in the object name `mh` stands for "minus half").

*Matrix
multiplication
operator %*%*

In order to perform the matrix multiplication in (A.4), the data frame `health.P` has to be converted to a regular matrix, and then the matrix multiplication is performed using the operator %*%, finally the SVD in (A.5) using function `svd()`:

```
> health.P <- as.matrix(health.P)
> health.S <- health.Drmh %*% (health.P - health.r %o% health.c)
+                             %*% health.Dcmh
> health.svd <- svd(health.S)
```

*Example of SVD
function* svd()

The principal and standard coordinates (`pc` and `sc`) are calculated as in (A.6–9):

```
> health.rsc <- health.Drmh
> health.csc <- health.Dcmh
> health.rpc <- health.rsc %*% diag(health.svd$d)
> health.cpc <- health.csc %*% diag(health.svd$d)
```

And that's it! The previous 14 R commands are the whole basic CA algorithm — simply replace `health` with any other data frame to compute the point coordinates.

To see the values of, for example, the principal coordinates of the rows on the first principal axis:

```
> health.rpc[,1]
[1] -0.37107411 -0.32988430 -0.19895401  0.07091332  0.39551813  ...
```

(notice that the signs are reversed compared to the display of Exhibit 6.3 — this can often occur using different software, but the user can reverse the signs of all the coordinates on an axis at will).

*Chapter 7:
Optimal Scaling*

Chapter 7 deals with optimal scaling properties of the CA solution, and there are no challenging calculations in this chapter. We illustrate the calculation of the transformed optimal scale values in (7.5) using R functions for calculating minimum and maximum values (again, because of the sign change in the coordinates on the first axis, the scale is reversed, in other words the transformed scale goes from 0=very good to 100=very bad; hence subtracting the results below from 100 will give the results of Exhibit 7.2):

```
> health.range <- max(health.csc[,1]) - min(health.csc[,1])
> health.scale <- (health.csc[,1] - min(health.csc[,1])) * 100 /
+                  health.range
> health.scale
[1]   0.00000  18.86467  72.42164  98.97005 100.00000
```

*Chapter 8:
Symmetry of Row
and Column
Analyses*

Chapter 8 is another chapter demonstrating properties of the solution rather than making new calculations. Exhibit 8.5 was not constructed using R but was typeset directly in LATEX (see descriptions of graphical typsetting at the end of this appendix). The maximal correlation properties of CA can be illustrated with some R commands, for example Equation (8.2) on page 63. The correlation between the age and health scale values on the first dimension is first calculated as $\phi^\mathsf{T}\mathbf{P}\gamma$ where ϕ and γ are the standard coordinates on the first dimension, and $\mathbf{P}$ is the correspondence matrix:

```
> health.cor <- t(health.rsc[,1]) %*% health.P %*% health.csc[,1]
> health.cor^2
          [,1]
[1,] 0.1366031
```

Thus the square of this correlation is the first principal inertia (the above result is given as a 1×1 matrix since it is the result of matrix-vector multiplications).

The following demonstrates the standardization in (A.12) for the rows, for example, which justifies the above way of calculating the correlation since the variances are 1:

```
> t(health.rsc[,1]) %*% health.Dr %*% health.rsc[,1]
      [,1]
[1,]    1
```

Chapter 9 explains the geometry of two-dimensional maps and compares asymmetric and symmetric maps. The smoking data set is part of the **ca** package so perhaps this is a good time to make a first introduction to that package. Once the package is installed and loaded, these data can be called up simply by issuing the command:

Chapter 9: Two-dimensional displays

```
> data(smoke)
```

which gives the data frame `smoke`:

Initial contact with **ca** *package*

```
> smoke
   none light medium heavy
SM    4     2      3     2
JM    4     3      7     4
SE   25    10     12     4
JE   18    24     33    13
SC   10     6      7     2
```

One of the functions in the **ca** package is `ca()` to perform simple CA. The CA of the smoking data is obtained easily by saying `ca(smoke)`:

```
> ca(smoke)

 Principal inertias (eigenvalues):
              1         2         3
 Value      0.074759 0.010017 0.000414
 Percentage 87.76%   11.76%    0.49%


 Rows:
                SM        JM        SE        JE        SC
 Mass      0.056995  0.093264  0.264249 0.455959  0.129534
 ChiDist   0.216559  0.356921  0.380779 0.240025  0.216169
 Inertia   0.002673  0.011881  0.038314 0.026269  0.006053
 Dim. 1   -0.240539  0.947105 -1.391973 0.851989 -0.735456
 Dim. 2   -1.935708 -2.430958 -0.106508 0.576944  0.788435
```

```
Columns:
              none     light    medium      heavy
Mass      0.316062 0.233161 0.321244  0.129534
ChiDist   0.394490 0.173996 0.198127  0.355109
Inertia   0.049186 0.007059 0.012610  0.016335
Dim. 1   -1.438471 0.363746 0.718017  1.074445
Dim. 2   -0.304659 1.409433 0.073528 -1.975960
```

Several numerical results are listed which should be familiar: the principal inertias and their percentages, and then for each row and column the mass, χ^2-distance to the centroid, the inertia, and the standard coordinates on the first two dimensions. The features of this package will be described in much more detail later on, but just to show now how simple the plotting is, simply put the **plot()** function around **ca(smoke)** to get the default symmetric CA map shown in Exhibit B.4:

```
> plot(ca(smoke))
```

Exhibit B.4:
*Symmetric map of
the data set* smoke,
using the **ca**
package.

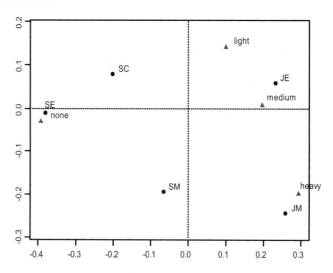

Notice that both principal axes have been inverted compared to the map of Exhibit 9.5. To obtain the asymmetric maps, add the option **map="rowprincipal"** or **map="colprincipal"** to the **plot()** function; for example, Exhibit 9.2 is obtained with the following command:

```
> plot(ca(smoke), map="rowprincipal")
```

*Chapter 10: Three
More Examples*

With this short introduction to the **ca** package, the analyses of Chapter 10 will be easy to reproduce. The three data sets are available on the website **www.carme-n.org** in text and Excel formats for copying and reading into R. The author data set is also provided with the **ca** package so this can be obtained, like the smoking data, with the R command **data(author)**. To see the author data in a three-dimensional CA map, try this (again, assuming you have loaded the **ca** package):

```
> data(author)
> plot3d.ca(ca(author), labels=c(2,1), sf=0.000001)
```

Chapter 11 involves a certain amount of new computations, all of which are actually part of the **ca** package, but here again we choose to demonstrate them first "by hand". The data set for the scientific research funding is read as described before — suppose the data frame is called fund. As in Chapter 4, the matrix of standardized residuals is calculated for this table, and then the inertias in Exhibit 11.1 are the sums-of-squares of the rows and columns:

Chapter 11:
Contributions to
Inertia

```
> fund.P <-as.matrix(fund / sum(fund))
> fund.r <-apply(fund.P, 1, sum)
> fund.c <-apply(fund.P, 2, sum)
> fund.Drmh <-diag(1 / sqrt(fund.r))
> fund.Dcmh <-diag(1 / sqrt(fund.c))
> fund.res <-fund.Drmh %*% (fund.P - fund.r %o% fund.c) %*% fund.Dcmh
> round(apply(fund.res^2, 1, sum), 5)
 [1] 0.01135 0.00990 0.00172 0.01909 0.01621 0.01256 0.00083
 [8] 0.00552 0.00102 0.00466
> round(apply(fund.res^2, 2, sum), 5)
[1] 0.01551 0.00911 0.00778 0.02877 0.02171
```

The permill contributions in Exhibit 11.2 are the squared standardized residuals relative to the total:

Contributions of
each cell of the
table to the total
inertia

```
> round(1000*fund.res^2 / sum(fund.res^2), 0)
      [,1] [,2] [,3] [,4] [,5]
 [1,]    0   32   16    0   89
 [2,]    0   23    4   44   48
 [3,]    3   12    1    0    5
 [4,]    9   15   11  189    8
 [5,]  106   11    2   74    3
 [6,]    1   11   38    1  102
 [7,]    2    0    0    3    5
 [8,]   51    4    0   10    2
 [9,]   10    0    0    2    0
[10,]    5    3   22   26    0
```

(the row and column labels have been lost because of the matrix multiplications, but can be restored again if necessary using rownames() and colnames() functions).

The principal inertias in Exhibit 11.3 are the squares of the singular values from the SVD of the residuals matrix:

```
> fund.svd <- svd(fund.res)
> fund.svd$d^2
[1] 3.911652e-02 3.038081e-02 1.086924e-02 2.512214e-03 3.793786e-33
```

(five values are given, but the fifth is theoretically an exact zero).

To calculate the individual components of inertia of the row points, say, on all four axes, we first need to calculate the principal coordinates f_{ik} (see (A.8)) and then the values of $r_i f_{ik}^2$:

```
> fund.F <- fund.Drmh %*% fund.svd$u %*% diag(fund.svd$d)
> fund.rowi <- diag(fund.r) %*% fund.F^2
> fund.rowi[,1:4]
              [,1]          [,2]          [,3]          [,4]
 [1,]  6.233139e-04  9.775878e-03  8.222230e-04  1.301601e-04
 [2,]  1.178980e-03  7.542243e-03  8.385857e-04  3.423076e-04
 [3,]  2.314352e-04  8.787604e-04  2.931994e-04  3.211261e-04
 [4,]  1.615600e-02  1.577160e-03  6.274587e-04  7.271264e-04
 [5,]  1.426048e-02  1.043783e-04  1.691831e-03  1.562740e-04
 [6,]  1.526183e-03  9.407586e-03  1.273528e-03  3.573707e-04
 [7,]  7.575664e-06  5.589276e-04  7.980532e-05  1.868385e-04
 [8,]  3.449918e-03  1.601539e-04  1.799425e-03  1.091335e-04
 [9,]  5.659639e-04  7.306881e-06  4.185906e-04  3.022249e-05
[10,]  1.116674e-03  3.684113e-04  3.024590e-03  1.516545e-04
```

which agrees with Exhibit 11.5. Notice in the last command above that only the first four columns are relevant (`fund.rowi[,1:4]`); there is a fifth column of tiny values which are theoretically zero because the fifth singular value is zero. Finally, to relativize these components with respect to the inertia of a point (row sums) or inertia of an axis (column sums, i.e., principal inertias) (see (A.27) and (A.26) respectively), and converting them to permills at the same time:

Calculating relative contributions (squared cosines or correlations)

```
> round(1000*(fund.rowi / apply(fund.rowi, 1, sum))[,1:4], 0)
      [,1] [,2] [,3] [,4]
 [1,]   55  861   72   11
 [2,]  119  762   85   35
 [3,]  134  510  170  186
 [4,]  846   83   33   38
 [5,]  880    6  104   10
 [6,]  121  749  101   28
 [7,]    9  671   96  224
 [8,]  625   29  326   20
 [9,]  554    7  410   30
[10,]  240   79  649   33
```

which agrees with Exhibit 11.6 (to obtain the qualities in Exhibit 11.8, add up the first two columns of the above table). With respect to column sums, i.e., principal inertias:

Calculating contributions to each principal axis

```
> round(1000*t(t(fund.rowi) / fund.svd$d^2)[,1:4], 0)
     [,1] [,2] [,3] [,4]
[1,]   16  322   76   52
[2,]   30  248   77  136
[3,]    6   29   27  128
[4,]  413   52   58  289
[5,]  365    3  156   62
```

```
[6,]    39  310  117  142
[7,]     0   18    7   74
[8,]    88    5  166   43
[9,]    14    0   39   12
[10,]   29   12  278   60
```

which shows how each axis is constructed; for example, rows 4 and 5 (Physics and Zoology) are the major contributors to the first axis.

Anticipating the fuller description of the **ca** package later, we point out that the complete set of these numerical results can be obtained using the summary() function around ca(fund) as follows:

```
> summary(ca(fund))
```

Principal inertias (eigenvalues):

```
dim    value      %     cum%   scree plot
1    0.039117  47.2   47.2   *************************
2    0.030381  36.7   83.9   *******************
3    0.010869  13.1   97.0   ******
4    0.002512   3.0  100.0
     --------  -----
Total: 0.082879 100.0
```

Rows:

	name	mass	qlt	inr	k=1	cor	ctr	k=2	cor	ctr	
1	Gel	107	916	137	76	55	16	303	861	322	
2	Bic	36	881	119	180	119	30	-455	762	248	
3	Chm	163	644	21	38	134	6	73	510	29	
4	Zol	151	929	230	-327	846	413	102	83	52	
5	Phy	143	886	196	316	880	365	27	6	3	
6	Eng	111	870	152	-117	121	39	-292	749	310	
7	Mcr	46	680	10	13	9	0	-110	671	18	
8	Bot	108	654	67	-179	625	88	-39	29	5	
9	Stt	36	561	12	125	554	14	14	7	0	
10	Mth	98	319	56	107	240	29	-61	79	12	

Columns:

	name	mass	qlt	inr	k=1	cor	ctr	k=2	cor	ctr	
1	A	39	587	187	478	574	228	72	13	7	
2	B	161	816	110	127	286	67	173	531	159	
3	C	389	465	94	83	341	68	50	124	32	
4	D	162	968	347	-390	859	632	139	109	103	
5	E	249	990	262	-32	12	6	-292	978	699	

Chapter 12: Supplementary Points

Chapter 12 shows how to add points to an existing map, using the barycentric relationship between standard coordinates of the column points, say, and the principal coordinates of the row points; i.e., profiles lie at weighted averages of vertices. The example at the top of page 94 shows how to situate the supplementary point *Museums* which has data [4 12 11 19 7] summing up to 53. Calculating the profile, say the vector **m** and then its scalar products with the standard column coordinates: $\mathbf{m}^\mathsf{T}\boldsymbol{\Gamma}$ gives its coordinates in the map:

```
> fund.m <- c(4,12,11,19,7)/53
> fund.Gamma <- fund.Dcmh%*%fund.svd$v
> t(fund.m) %*% fund.Gamma[,1:2]
          [,1]        [,2]
[1,] -0.3143203 0.3809511
```

(the sign of the second axis is reversed in this solution compared to Exhibit 12.2). It is clear that if we perform the same operation with the unit vectors of Exhibit 12.4 as supplementary points, then multiplying these with the standard coordinates is just the same as the standard coordinates.

Chapter 13: Correspondence Analysis Biplots

In Chapter 13 the different scaling of a CA map are discussed from the point of view of the biplot. In the standard CA biplot of Exhibit 13.3 the rows are in principal coordinates while the columns are in rescaled standard coordinates where each column point has been pulled in by multiplying its coordinates by the square root of the column mass. Given the standard coordinates fund.Gamma calculated above, these rescaled coordinates on the first two dimensions are calculated as:

```
> diag(sqrt(fund.c)) %*% fund.Gamma[,1:2]
          [,1]         [,2]
[1,]  0.47707276  0.08183444
[2,]  0.25800640  0.39890356
[3,]  0.26032157  0.17838093
[4,] -0.79472740  0.32170520
[5,] -0.08046934 -0.83598151
```

In the following commands, the scalar products on the right-hand side of (13.7), for $K^* = 2$, are first stored in fund.est and then estimated profiles are calculated by multiplying by the square roots $\sqrt{c_j}$ and adding c_j, all using matrix algebra:

```
> fund.est <- fund.F[,1:2] %*% t(diag(sqrt(fund.c)) %*%
+             fund.Gamma[,1:2])
> oner <- rep(1,dim(fund)[1])
> round(fund.est %*% diag(sqrt(fund.c)) + oner %o% fund.c, 3)
          A     B     C     D     E
 [1,] 0.051 0.217 0.436 0.177 0.120
 [2,] 0.049 0.107 0.368 0.046 0.431
 [3,] 0.044 0.176 0.404 0.160 0.217
 [4,] 0.010 0.143 0.348 0.280 0.219
 [5,] 0.069 0.198 0.444 0.065 0.225
 [6,] 0.023 0.102 0.338 0.162 0.375
 [7,] 0.038 0.145 0.379 0.144 0.294
 [8,] 0.021 0.136 0.356 0.214 0.272
 [9,] 0.051 0.176 0.411 0.124 0.238
[10,] 0.048 0.162 0.400 0.120 0.270
```

This can be compared with the true profile values:

```
> round(fund.P/fund.r, 3)
```

```
        A     B     C     D     E
Geol 0.035 0.224 0.459 0.165 0.118
Bioc 0.034 0.069 0.448 0.034 0.414
Chem 0.046 0.192 0.377 0.162 0.223
Zool 0.025 0.125 0.342 0.292 0.217
Phys 0.088 0.193 0.412 0.079 0.228
Engi 0.034 0.125 0.284 0.170 0.386
Micr 0.027 0.162 0.378 0.135 0.297
Bota 0.000 0.140 0.395 0.198 0.267
Stat 0.069 0.172 0.379 0.138 0.241
Math 0.026 0.141 0.474 0.103 0.256
```

Calculating the differences between the true and estimated profile values gives the individual errors of approximation, and the sum of squares of these differences, suitably weighted, gives the overall error in the two-dimensional CA. Each row of squared differences has to be weighted by the corresponding row mass r_i and each column by the inverse of the expected value $1/c_j$. The calculation is the following (this is one command, wrapped over two lines here, a concentrated example in R programming!):

```
> sum(diag(fund.r) %*% (fund.est%*%diag(sqrt(fund.c))+
+   oner %o% fund.c - fund.P / fund.r)^2 %*% diag(1/fund.c))
[1] 0.01338145
```

To demonstrate that this is correct, add the principal inertias *not* on the first two axes:

```
> sum(fund.svd$d[3:4]^2)
[1] 0.01338145
```

which confirms the previous calculation (this is the 16% unexplained inertia reported at the bottom of page 103).

The calculation of the biplot calibrations is quite intricate since it involves a lot of trigonometry. Rather than list the whole procedure here, the interested reader is referred to the website where the script for the function `biplot.ca` is given and which calculates the coordinates of the starting and ending points of all the tic marks on the biplot axes for the columns.

Biplot axis calibration

In Chapter 14 various linear relationships between row and column coordinates and the data are given. Here we shall demonstrate some of these using R's linear modelling function `lm()` which allows weights to be specified in the least-squares regression. For example, let's perform the weighted least-squares regression of the standard row coordinates (y-axis in Exhibit 14.2) on the column standard coordinates (x-axis). The variables of the regression have 10×5 values, and these will be vectorized in columns corresponding to the original matrix. Thus the x variable is the vector (called `fund.vecc` below) where the first column coordinate on the first dimension is repeated 10 times, then the second coordinate 10 times and so on, whereas the y variable (`fund.vecr`) has the set of first dimension's row coordinates repeated five times in a column (the row standard coordinates are calculated as `fund.Phi` below). Check the

Chapter 14: Transition and Regression Relationships

values of `fund.vecc` and `fund.vecr` below as you perform the computations. The weights of the regression will be the frequencies in the original table `fund` — to vectorize these, the data frame has to be first converted to a matrix and then to a vector using `as.vector()`:

Conversion of data objects using `as.matrix()` *and* `as.vector()`

```
> fund.vec <- as.vector(as.matrix(fund))
> fund.Phi <- fund.Drmh %*% fund.svd$u
> fund.vecr <- rep(fund.Phi[,1], 5)
> fund.vecc <- as.vector(oner %*% t(fund.Gamma[,1]))
```

The weighted least-squares regression is then performed as follows:

Example of `lm()` *function for linear regression, using* `weights` *option*

```
> lm(fund.vecr~fund.vecc, weight = fund.vec)

Call:
lm(formula = fund.vecr ~ fund.vecc, weights=fund.vec)

Coefficients:
(Intercept)      fund.vecc
 -4.906e-16      1.978e-01
```

showing that the constant is zero and the coefficient is 0.1978, the square root of the first principal inertia.

To perform the regression described on page 110 between Geology's contingency ratios and the standard coordinates on the first two dimensions, the response `fund.y` is regressed on the first two columns of the standard coordinate matrix Γ in `fund.Gamma`, with weights c in `fund.c`, as follows (here the `summary()` function is used around the `lm` command to get more results):

```
> fund.y <- (fund.P[1,] / fund.r[1]) / fund.c
> summary(lm(fund.y ~ fund.Gamma[,1] + fund.Gamma[,2],
+              weights=fund.c))

Call:
lm(formula = fund.y ~ fund.Gamma[, 1]+fund.Gamma[, 2], weights=fund.c)

Residuals:
         A          B          C          D          E
 -0.079708   0.016013   0.037308  -0.030048  -0.003764

Coefficients:
                  Estimate Std. Error t value Pr(>|t|)
(Intercept)        1.00000    0.06678  14.975  0.00443 **
fund.Gamma[, 1]    0.07640    0.06678   1.144  0.37105
fund.Gamma[, 2]    0.30257    0.06678   4.531  0.04542 *
---
Signif. codes:  0 '***' 0.001 '**' 0.01 '*' 0.05 '.' 0.1 ' ' 1

Residual standard error: 0.06678 on 2 degrees of freedom
Multiple R-Squared: 0.9161,      Adjusted R-squared: 0.8322
F-statistic: 10.92 on 2 and 2 DF,  p-value: 0.0839
```

confirming the coefficients at the bottom of page 110 (again, the second coefficient has reversed sign because the second dimension coordinates are reversed) and the R^2 of 0.916.

The lm() function does not give standardized regression coefficients, but these can be obtained by calculating weighted correlations using the weighted covariance function cov.wt() with option cor=TRUE:

```
> cov.wt(cbind(fund.y,fund.Gamma[,1:2]), wt=fund.c, cor=TRUE)$cor
$cor
            [,1]         [,2]         [,3]
[1,] 1.0000000 2.343286e-01 9.280040e-01
[2,] 0.2343286 1.000000e+00 2.359224e-16
[3,] 0.9280040 2.359224e-16 1.000000e+00
```

Example of cov.wt *function to calculate weighted correlation*

which agrees with the correlation matrix at the top of page 111, apart from possible sign changes.

Chapter 15 deals with Ward clustering of the row or column profiles using Ward clustering, weighting the profiles by their masses. The R function for performing hierarchical clustering is hclust(), which does not allow differential weights in the option for Ward clustering (see (15.2)); neither does the function agnes() in the package **cluster**. The commercial statistical package XLSTAT, described later, does have this possibility. In addition, Fionn Murtagh's R programs also include Ward clustering with weights (see page 253).

Chapter 15: Clustering the Rows and Columns

In Chapter 16 the interactive coding of variables was described. To be able to code the data in this way, either the multiway table is needed or the original data. For example, in the case of the health data used in Chapter 16, the raw data looks like this, showing the first four rows of data out of 6371):

Chapter 16: Multiway Tables

```
. . .    health     age      gender  . . .
. . .       4         5         2     . . .
. . .       2         3         1     . . .
. . .       2         4         1     . . .
. . .       3         5         1     . . .
. . .       .         .         .     . . .
. . .       .         .         .     . . .
```

To obtain Exhibit 16.2, the seven categories of age and the two categories of gender need to be combined into one variable age_gender with 14 categories. This is achieved with a simple transformation such as:

```
> age_gender <- 7 * (gender - 1) + age
```

Interactive coding

which will make the age groups of male (gender=1) numbered 1 to 7, and those of female (gender=2) numbered 8 to 14. From then on, everything continues as before, with a cross-tabulation being made of the variable age_gender with health. Cross-tabulations in R are made with the function table(); for example,

Cross-tabulations with table()

```
> table(age_gender, health)
```

would give the cross-tabulation in Exhibit 16.2.

Now suppose the raw data from the data set on working women is in an Excel file as shown below: four questions from Q1 to Q4, country (C), gender (G), age (A), marital status (M) and education (E). To input the data into R, copy the

columns to the clipboard as before, using function read.table(). But now the table does not have row names, and so no blank in the top left-hand cell; hence the option header=T needs to be specified (T is short for TRUE):

```
> women <- read.table("clipboard", header=T)
```

The column names of data frame women are obtained using function colnames:

```
> colnames(women)
 [1] "Q1" "Q2" "Q3" "Q4" "C"  "G"  "A"  "M"  "E"
```

Example of **attach()** *function*

In order to obtain the table in Exhibit 16.4, it is convenient to use the attach() function, which allows all the column names listed above to be available as if they were regular object names (to make these names unavailable the inverse operation detach() should be used):

```
> attach(women)
> table(C, Q3)
    Q3
C      1    2    3    4
  1  256 1156  176  191
  2  101 1394  581  248
  3  278  691   62   66
  4  161  646   70  107

  .    .    .    .    .

  .    .    .    .    .
 21  243  448  484   25
 22  468  664   92   63
 23  203  671  313  120
 24  738 1012  514  230
```

(cf. Exhibit 16.4).

To get the interactively coded row variable in Exhibit 16.6 and the table itself:

```
> CG <- 2 * (C - 1) + G
> table(CG, Q3)
     Q3
CG     1   2   3   4
  1  117 596 114  82
  2  138 559  60 109
  3   43 675 357 123
  4   58 719 224 125
  .    .   .   .   .

  .    .   .   .   .
 47  348 445 294 112
 48  390 566 218 118
 51    1   2   0   0
 55    1   1   2   1
```

Notice that the last two rows of the table correspond to a few missing values for gender that were coded as 9; to see frequency counts for each column, enter the command `lapply(women,table)`. So we should remove all the missing data first — see page 237 how to remove cases with missing values. Alternatively, missing values can be assigned R's missing value code NA, for example the missing values for gender in column 6:

```
> women[,6][G==9]<-NA
> attach(women)
> CG <- 2 * (C - 1) + G
```

(notice that data frame `women` has to be attached again and CG recomputed). Assuming that all missing values have been recoded (or cases removed), the combinations of CG and A are coded as follows in order to construct the variable with 288 categories that interactively codes country, gender and age group (there are no missings for age):

```
> CGA <- 6 * (CG - 1) + A
```

Chapter 17: Stacked Tables

Chapter 17 considers the CA of several cross-tables concatenated (or juxtaposed). The two functions `rbind()` and `cbind()` provide the tools for binding rows or columns together. For example, assuming the 33590×10 raw data matrix `women` is available and attached as described above, then the five cross-tabulations corresponding to Question 3 depicted in Exhibit 17.1 can be built up as follows in a `for` loop:

```
> women.stack<-table(C, Q3)
> for(j in 6:9){
+   women.stack <- rbind(women.stack, table(women[,j], Q3))
+ }
```

Notice how the columns of `women` can be accessed by name or by column number. If you look at the contents of `women.stack` you will see several rows corresponding to missing data codes for all demographic variables except country and age group. These would have to be omitted before the CA is performed, which can be done in three different ways: (i) by excluding these rows from the matrix, e.g., if rows 38, 39, 47 and 48 correspond to missing values, then remove as follows:

```
> women.stack <- women.stack[-c(38,39,47,48),]
```

(the negative sign before the set of row numbers indicates exclusion); (ii) by changing the missing value codes to NAs as described on the previous page; or (iii) by declaring the missing rows outside the subset of interest in a subset CA as described in Chapter 21 (this is the best option, since it keeps the sample size in each table the same).

χ^2 *statistic using* χ^2 *test function* `chisq.test()`

To check the inertias in the table on page 124, we can try R's χ^2 test function `chisq.test` which has as one of its results the χ^2 statistic specified by `$statistic`. We make the calculation for the age variable's cross-tabulation with question 3, which corresponds to rows 27 to 32 of the stacked matrix (after 24 rows for country and 2 rows for gender). Dividing the statistic by the sample size, the total of the table, gives the inertia:

```
> chisq.test(women.stack[27:32,])$statistic /
+                 sum(women.stack[27:32,])

X-squared
0.0421549
```

which agrees with the value for age in the table on page 124.

To build up the table in Exhibit 17.4, the four stacked tables (each with five tables) for the four questions are column-bound using `cbind()`.

The **ca** *package*

At this point, before we enter the intricacies of MCA and its related methods, we are going to leave the "by hand" R exercizes behind and start to use the functions in the **ca** package routinely. The package comprises functions for simple, multiple and joint CA with support for subset analyses and the inclusion of supplementary variables. Furthermore, it offers functions for the graphical display of the results in two and three dimensions. The package is comprised of the following components:

- Simple CA:
  - Computation: `ca()`
  - Printing and summaries: `print.ca()` and `summary.ca()` (and `print.summary.ca()`)
  - Plotting: `plot.ca()` and `plot3d.ca()`
- MCA and JCA:
  - Computation: `mjca()`
  - Printing and summaries: `print.mjca()` and `summary.mjca()` (and `print.summary.mjca()`)
  - Plotting: `plot.mjca()` and `plot3d.mjca()`
- Data sets:
  - `smoke`, `author` and `wg93`

The package contains further functions, such as `iterate.mjca()` for the updating of the Burt matrix in JCA.

The function `ca()` computes simple CA, for example

ca() *function*

```
> library(ca)    #this loads ca if not already done using R menu
> data(smoke)
> ca(smoke)
```

performs a simple CA on the `smoke` data set (see pages 221–222). A list of all available entries that are returned by `ca()` is obtained with `names()`:

```
> names(ca(smoke))
 [1] "sv"         "nd"          "rownames"    "rowmass"    "rowdist"
 [6] "rowinertia" "rowcoord"    "rowsup"      "colnames"   "colmass"
[11] "coldist"    "colinertia"  "colcoord"    "colsup"     "call"
```

The output of `ca()` is structured as a list-object; for example, the row standard coordinates are obtained with

```
> ca(smoke)$rowcoord
```

Optional arguments for the `ca()` function include an option for setting the dimensionality of the solution (`nd`), options for marking selected rows and/or columns as supplementary ones (`suprow` and `supcol`, respectively) and options for setting subset rows and/or columns (`subsetrow` and `subsetcol`, respectively) for subset CA.

As an extension to the printing method, a summary method is also provided. This gives a more detailed output as follows:

```
> summary(ca(smoke))
```

returns the summary of the CA:

```
Principal inertias (eigenvalues):
 dim    value     %    cum%  scree plot
  1    0.074759  87.8  87.8  ************************
  2    0.010017  11.8  99.6  ***
  3    0.000414   0.5 100.0
       --------  -----
Total: 0.085190 100.0


Rows:
      name   mass  qlt   inr    k=1 cor ctr    k=2 cor ctr
 1 |    SM |   57  893    31 |  -66  92   3 | -194 800 214 |
 2 |    JM |   93  991   139 |  259 526  84 | -243 465 551 |
 3 |    SE |  264 1000   450 | -381 999 512 |  -11   1   3 |
 4 |    JE |  456 1000   308 |  233 942 331 |   58  58 152 |
 5 |    SC |  130  999    71 | -201 865  70 |   79 133  81 |


Columns:
      name   mass  qlt   inr    k=1 cor ctr    k=2 cor ctr
 1 |   non |  316 1000   577 | -393 994 654 |  -30   6  29 |
 2 |   lgh |  233  984    83 |   99 327  31 |  141 657 463 |
 3 |   mdm |  321  983   148 |  196 982 166 |    7   1   2 |
 4 |   hvy |  130  995   192 |  294 684 150 | -198 310 506 |
```

Again, eigenvalues and relative percentages of explained inertia are given for all available dimensions. Additionally, cumulated percentages and a scree plot are shown. The items given in `Rows` and `Columns` include the principal coordinates for the first two dimensions (k=1 and k=2). Squared correlation (`cor`) and contributions (`ctr`) for the points are displayed next to the coordinates. The quantities in these tables are multiplied by 1000 (e.g., the coordinates and masses), which for `cor` and `ctr` means that they are expressed in thousandths, or permills (‰). Quality (`qlt`) is given for the requested solution; i.e., in this case it is the sum of the squared correlations for the first two dimensions. In the case of supplementary variables, an asterisk is appended to the variable names in the output; for example, the summary for the CA of the `smoke` data, where the `none` category (the first column) is treated as supplementary, is:

```
> summary(ca(smoke, supcol=1))
```

In the corresponding section of the output the following is given:

```
...
Columns:
       name   mass  qlt  inr    k=1 cor  ctr    k=2 cor  ctr
  1 | (*)non | <NA>   55 <NA> |  292  39 <NA> | -187  16 <NA> |
...
```

showing that masses, inertias and contributions are "not applicable".

Graphical displays in the **ca** *package*   The graphical representation of CA and MCA solutions is commonly done with *symmetric* maps, and this is with the default option in the `plot()` function (`map="symmetric"`). The complete set of `map` options is as follows:

—	`"symmetric"`	Rows and columns in principal coords (default) i.e., scaled to have inertia equal to principal inertia (eigenvalue, or square of singular value)
—	`"rowprincipal"`	Rows in principal and columns in standard coords
—	`"colprincipal"`	Columns in principal and rows in standard coords
—	`"symbiplot"`	Row and column coords are scaled to have inertias equal to the singular values
—	`"rowgab"`	Rows in principal coords and columns in standard coords times mass (according to a proposal by Gabriel)
—	`"colgab"`	Columns in principal coords and rows in standard coords times mass
—	`"rowgreen"`	Rows in principal coords and columns in standard coords times square root of mass (according to a proposal by Greenacre — see Chapter 13)
—	`"colgreen"`	Columns in principal coords and rows in standard coords times square root of mass

By default, supplementary variables are added to the plot with a different symbol. The symbols can be defined with the `pch` option in `plot.ca()`. This option takes four values in the following order: plotting point character or

symbol for (i) active rows, (ii) supplementary rows, (iii) active columns and
(iv) supplementary columns. As a general rule, options that contain entries
for rows and for columns contain the entries for the rows first and then those
for the columns. For example, the colour of the symbols is specified with the
col option; by default it is `col=c("#000000", "#FF0000")` — black for rows
and red for columns. Instead of these hexadecimal codes, there is a reduced
list with names such as `"black"`, `"red"`, `"blue"`, `"green"`, `"gray"`, etc.

The option `what` controls the content of the plot. It can be set to `"all"`,
`"active"`, `"passive"` or `"none"` for the rows and for the columns. For example, a plot of only the active (i.e., excluding supplementary) points is created
by using `what=c("active", "active")`.

In addition to the `map` scaling options, various options allow certain values
to be added to the plot as graphical attributes. The option `mass` selects if
the masses of rows or columns should be indicated by the size of the point.
Similarly, relative or absolute contributions can be indicated by the colour
intensity in the plot by using the `contrib` option.

The option `dim` selects which dimensions to plot, the default being `dim=c(1,2)`,
i.e., the first two dimensions are plotted. A plot of the second and third dimensions, for example, is obtained by setting `dim=c(2,3)`. Another possibility for adding the third dimension to the plot is given with the functions
`plot3d.ca()` and `plot3d.mjca()`. These two functions rely on the **rgl** package for three-dimensional graphics in R. Their structure is kept similar to their
counterparts for two dimensions; for example,

```
> plot3d(ca(smoke, nd=3))
```

creates a three-dimensional display of the CA, shown in Exhibit B.5.

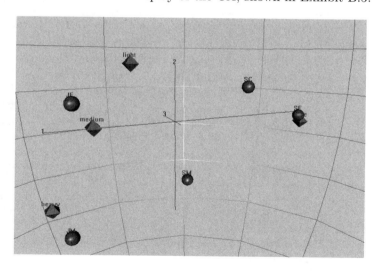

Exhibit B.5:
*Three-dimensional
display of a simple
CA (compare with
two-dimensional
map in Exhibit B.4).*

This display can be rotated and also zoomed in or out using the mouse and
its buttons.

MCA and JCA are performed with the function `mjca()`. The structure of the function is kept similar to its counterpart from simple CA. The two most striking differences are the format of the input data and the restriction to columns for the analyses. The function `mjca()` takes a response pattern matrix as input.

Within the function, the response pattern matrix is converted to an indicator matrix and a Burt matrix, depending on the type of analysis. Restricting to columns means that only values for the columns are given in the output and the specification of supplementary variables is limited to columns. The "approach" to MCA is specified by the `lambda` option in `mjca()`:

- `lambda="indicator"`: Analysis based on a simple CA of the indicator matrix

- `lambda="Burt"`: Analysis based on an eigenvalue-decomposition of the Burt matrix

- `lambda="adjusted"`: Analysis based on the Burt matrix with an adjustment of inertias (default)

- `lambda="JCA"`: Joint correspondence analysis

By default, `mjca()` performs an adjusted analysis, i.e., `lambda="adjusted"`. For JCA (`lambda="JCA"`), the Burt matrix is updated iteratively by weighted least squares, using the internal function `iterate.mjca()`. This updating function has two convergence criteria, namely `epsilon` and `maxit`. Option `epsilon` sets a convergence criterion by means of maximum absolute difference of the Burt matrix in an iteration step compared to the Burt matrix of the previous step. The maximum number of iterations is given by the option `maxit`. The program iterates until any one of the two conditions is satisfied. Setting one option to `NA` results in ignoring that criterion; for example, exactly 50 iterations without considering convergence are performed with `maxit=50` and `epsilon=NA`.

As with simple CA, the solution is restricted by the `nd` option to two dimensions. However, eigenvalues are given for all possible dimensions, which number $(J - Q)$ for the "indicator" and "Burt" versions of MCA. In the case of an adjusted analysis or a JCA, the eigenvalues are given only for those dimensions k, where the singular values from the Burt matrix λ_k (i.e., the principal inertias of the indicator matrix) satisfy the condition $\lambda_k > 1/Q$.

In Chapter 18 the data on working women, for West and East German samples, were analysed using the indicator and Burt versions of MCA. Assuming that the `women` data frame (with 33590 rows) read previously is available and "attached", the two German samples have country codes 2 and 3 repectively. The part of the `women` corresponding to these two samples can be accessed using a *logical* vector which we call `germany`:

```
> germany <- C==2 | C==3
> womenG  <- women[germany,]
```

The first command creates a vector of length 33590 with values `TRUE` corresponding to the rows of the German samples, otherwise `FALSE`. The second command then passes only those rows with `TRUE` values to the new data frame `womenG`. There are 3421 rows in `womenG`, whereas the matrix analysed in Chapter 18 has 3418 rows — three cases that have some missing demographic data have been eliminated (i.e., *listwise deletion* of missings). Variables gender, marital status and education have missing value codes 9, except for education where they are 98 and 99. The steps needed to eliminate the missing rows use the same method as above to flag the rows and then eliminate them:

```
> missing <- G==9 | M==9 | E==98 | E==99
> womenG  <- (womenG[!missing,])
```

*Listwise deletion
of missing values*

(If missing values have been replaced by R's `NA` code, as described on page 231, then use `NA` in the above.)

The indicator version of MCA for the first four columns (the four questions on women working or staying at home) is obtained simply as follows:

```
> mjca(womenG[,1:4], lambda="indicator")
```

Eigenvalues:

	1	2	3	4	5	6
Value	0.693361	0.513203	0.364697	0.307406	0.21761	0.181521
Percentage	23.11%	17.11%	12.16%	10.25%	7.25%	6.05%
	7	8	9	10	11	12
Value	0.164774	0.142999	0.136322	0.113656	0.100483	0.063969
Percentage	5.49%	4.77%	4.54%	3.79%	3.35%	2.13%

Columns:

	Q1.1	Q1.2	Q1.3	Q1.4	Q2.1	Q2.2	
Mass	0.182929	0.034816	0.005778	0.026477	0.013239	0.095012	...
ChiDist	0.605519	2.486096	6.501217	2.905510	4.228945	1.277206	...
Inertia	0.067071	0.215184	0.244222	0.223523	0.236761	0.154988	...
Dim. 1	-0.355941	-0.244454	-0.279167	2.841498	-0.696550	-0.428535	...
Dim. 2	-0.402501	1.565682	3.971577	-0.144653	-2.116572	-0.800930	...

and the Burt version:

```
> mjca(womenG[,1:4], lambda="Burt")
```

Eigenvalues:

	1	2	3	4	5	6
Value	0.480749	0.263377	0.133004	0.094498	0.047354	0.03295
Percentage	41.98%	23%	11.61%	8.25%	4.13%	2.88%
	7	8	9	10	11	12
Value	0.027151	0.020449	0.018584	0.012918	0.010097	0.004092
Percentage	2.37%	1.79%	1.62%	1.13%	0.88%	0.36%

Columns:

	Q1.1	Q1.2	Q1.3	Q1.4	Q2.1	Q2.2	
Mass	0.182929	0.034816	0.005778	0.026477	0.013239	0.095012	...
ChiDist	0.374189	1.356308	3.632489	2.051660	2.354042	0.721971	...
Inertia	0.025613	0.064046	0.076244	0.111452	0.073363	0.049524	...
Dim. 1	0.355941	0.244454	0.279167	-2.841498	0.696550	0.428535	...

```
Dim. 2  -0.402501 1.565682 3.971577 -0.144653 -2.116572 -0.800930 ...
```

The total inertia can be computed in the two cases as the sum of squared singular values, as in the simple CA case:

```
> sum(mjca(womenG[,1:4], lambda="indicator")$sv^2)
[1] 3

> sum(mjca(womenG[,1:4], lambda="Burt")$sv^2)
[1] 1.145222
```

The contributions of each subtable of the Burt matrix to the total inertia is given in the component called `subinertia` of the `mjca` object, so the sum of these also gives the total inertia:

```
> sum(mjca(womenG[,1:4], lambda="Burt")$subinertia)
[1] 1.145222
```

Since the total inertia is the average of the 16 subtables, the inertia of individual subtables are 16 times the values in $subinertia:

```
> 16*mjca(womenG[,1:4], lambda="Burt")$subinertia
           [,1]        [,2]        [,3]        [,4]
[1,] 3.0000000 0.3657367 0.4261892 0.6457493
[2,] 0.3657367 3.0000000 0.8941517 0.3476508
[3,] 0.4261892 0.8941517 3.0000000 0.4822995
[4,] 0.6457493 0.3476508 0.4822995 3.0000000
```

To obtain the positions of the supplementary variables:

```
> summary(mjca(womenG, lambda="Burt", supcol=5:9))

Principal inertias (eigenvalues):

 dim    value      %     cum%  scree plot
  1    0.480749   42.0   42.0  *************************
  2    0.263377   23.0   65.0  **************
  3    0.133004   11.6   76.6  *******
  4    0.094498    8.3   84.8  *****
  5    0.047354    4.1   89.0  **
  6    0.032950    2.9   91.9  **
  7    0.027151    2.4   94.2  *
  8    0.020449    1.8   96.0  *
  9    0.018584    1.6   97.6  *
 10    0.012918    1.1   98.8
 11    0.010097    0.9   99.6
 12    0.004092    0.4  100.0
       --------  -----
Total: 1.145222  100.0


Columns:
          name  mass  qlt  inr  |   k=1   cor  ctr  |   k=2  cor  ctr  |
 1  |  Q1.1  |  183   740    6  |   247   435   23  |  -207  305   30  |
 2  |  Q1.2  |   35   367   14  |   169    16    2  |   804  351   85  |
 3  |  Q1.3  |    6   318   16  |   194     3    0  |  2038  315   91  |
 4  |  Q1.4  |   26   923   24  | -1970   922  214  |   -74    1    1  |
```

```
5  |   Q2.1  |   13  255   16  |   483   42    6 | -1086  213   59 |
6  |   Q2.2  |   95  494   11  |   297  169   17 |  -411  324   61 |
.  |    .    |    .    .    .  |    .     .    . |    .     .    . |
.  |    .    |    .    .    .  |    .     .    . |    .     .    . |
17 | (*)C.2  | <NA>  283 <NA>  |   -89   48 <NA> |   195  234 <NA> |
18 | (*)C.3  | <NA>  474 <NA>  |   188   81 <NA> |  -413  393 <NA> |
19 | (*)G.1  | <NA>   26 <NA>  |   -33    5 <NA> |    67   21 <NA> |
20 | (*)G.2  | <NA>   24 <NA>  |    34    5 <NA> |   -68   19 <NA> |
21 | (*)A.1  | <NA>   41 <NA>  |  -108   12 <NA> |  -170   29 <NA> |
22 | (*)A.2  | <NA>   52 <NA>  |   -14    0 <NA> |  -172   52 <NA> |
.  |    .    |    .    .    .  |    .     .    . |    .     .    . |
.  |    .    |    .    .    .  |    .     .    . |    .     .    . |
```

The supplementary categories are marked by a * and have no masses, inertia values (`inr`) nor contributions to the principal axes (`ctr`).

To obtain the JCA of the same data as in Exhibit 19.3, simply change the `lambda` option to `"JCA"`. Here percentages of inertia are not given for individual axes, but only for the solution space as a whole, since the axes are not nested:

Chapter 19: Joint Correspondence Analysis

```
> summary(mjca(womenG[,1:4], lambda="JCA"))

Principal inertias (eigenvalues):

1      0.353452
2      0.128616
3      0.015652
4      0.003935
       --------
Total: 0.520617


Diagonal inertia discounted from eigenvalues: 0.125395
Percentage explained by JCA in 2 dimensions: 90.2%
(Eigenvalues are not nested)
[Iterations in JCA: 31 , epsilon = 9.33e-05]

Columns:
      name   mass  qlt  inr     k=1  cor  ctr      k=2  cor  ctr
1  | Q1.1 |   183  969   21 |    204  693   22 |   -129  276   24 |
2  | Q1.2 |    35  803   23 |    144   61    2 |    503  742   69 |
3  | Q1.3 |     6  557   32 |    163    9    0 |   1260  548   71 |
4  | Q1.4 |    26  992  137 |  -1637  991  201 |    -45    1    0 |
5  | Q2.1 |    13  597   31 |    394  125    6 |   -764  471   60 |
6  | Q2.2 |    95  956   26 |    250  431   17 |   -276  525   56 |
.  |   .  |    .    .    . |     .    .    . |      .    .    . |
.  |   .  |    .    .    . |     .    .    . |      .    .    . |
```

The squared correlations, and thus the qualities too, are all much higher in the JCA.

Notice that in the JCA solution, the "total" inertia is the inertia of the modified Burt matrix, which includes a part due to the modified diagonal blocks — this additional part is the "`Diagonal inertia discounted from eigenvalues: 0.125395`" which has to be subtracted from the total to get the total inertia due to the off-diagonal blocks. Since the solution requested is two-dimensional and fits the diagonal blocks exactly by construction, the first two eigenvalues also contain this additional part, which has to be discounted as well. The proportion of (off-diagonal) inertia explained is thus:

$$\frac{0.3534 + 0.1286 - 0.1254}{0.5206 - 0.1254} = 0.9024$$

i.e., the percentage of 90.2% reported above (see Theoretical Appendix, (A.32)). The denominator above, the adjusted total $0.5206 - 0.1254 = 0.3952$, can be verified to be the same as:

$$\text{inertia of } \mathbf{B} - \frac{J - Q}{Q} = 1.1452 - \frac{12}{16} = 0.3952$$

To obtain the adjusted MCA solution, that is the same standard coordinates as in MCA but (almost) optimal scaling factors ("almost" optimal because the nesting property is retained, whereas the optimal adjustments do not preserve nesting), either use the `lambda` option `"adjusted"` or leave out this option since it is the default:

```
> summary(mjca(womenG[,1:4]))

Principal inertias (eigenvalues):
```

```
dim    value     %    cum%  scree plot
1     0.349456  66.3  66.3  ************************
2     0.123157  23.4  89.7  *********
3     0.023387   4.4  94.1  *
4     0.005859   1.1  95.2
```

```
Adjusted total inertia: 0.526963
```

```
Columns:
       name   mass  qlt  inr     k=1  cor ctr     k=2  cor ctr
1  | Q1.1 |   183  996   22 |    210  687  23 |   -141  309  30 |
2  | Q1.2 |    35  822   26 |    145   53   2 |    549  769  85 |
3  | Q1.3 |     6  562   38 |    165    8   0 |   1394  554  91 |
4  | Q1.4 |    26 1009  141 |  -1680 1008 214 |    -51    1   1 |
5  | Q2.1 |    13  505   36 |    412  119   6 |   -743  387  59 |
6  | Q2.2 |    95  947   27 |    253  424  17 |   -281  522  61 |
   .    .     .    .    .       .    .    .       .    .    .
   .    .     .    .    .       .    .    .       .    .    .
```

The adjusted total inertia, used to calculate the percentages above, is calculated just after (19.5) on page 149. The first two adjusted principal inertias (eigenvalues) are calculated just after (19.6) (see also (A.35) and (A.36)).

Chapter 20:
Scaling Properties
of MCA

Chapter 20 generalizes the ideas of Chapter 7, also Chapter 8, to the multivariate case. The data set used in this chapter is the science and environment data available as a data set in our **ca** package, so all you need to do to load this data set is to issue the command:

```
> data(wg93)
```

The resulting data frame `wg93` contains the four questions described on page 153, as well as three demographic variables: gender, age and education (the last two have six categories each). The MCA map of Exhibit 20.1 is obtained as follows, this time after saving the MCA results in object `wg93.mca`:

```
> wg93.mca <- mjca(wg93[,1:4], lambda="indicator")
> plot(wg93.mca, what=c("none", "all"))
```

The map might turn out inverted on the first or second axis, but that is — as we have said before — of no consequence.

Exhibit 20.2 is obtained by formatting the contributions to axis 1 as a 5×4 matrix (first the principal coordinates `wg93.F` are calculated, then raw contributions `wg93.coli`):

```
> wg93.F    <- wg93.mca$colcoord %*% sqrt(wg93.mca$sv)
> wg93.coli <- diag(wg93.mca$colmass)%*%wg93.F^2
> matrix(round(1000*wg93.coli[,1] / wg93.mca$sv[1]^2, 0), nrow=5)
     [,1] [,2] [,3] [,4]
[1,]  115  174  203   25
[2,]   28   21    6    3
[3,]   12    7   22    9
[4,]   69   41   80    3
[5,]   55   74   32   22
```

The following commands assign the first standard coordinates as the four item scores for each of the 871 respondents, and the average score:

```
> Ascal <- wg93.mca$colcoord[1:5,1]
> Bscal <- wg93.mca$colcoord[6:10,1]
> Cscal <- wg93.mca$colcoord[11:15,1]
> Dscal <- wg93.mca$colcoord[16:20,1]
> As    <- Ascal[wg93[,1]]
> Bs    <- Bscal[wg93[,2]]
> Cs    <- Cscal[wg93[,3]]
> Ds    <- Dscal[wg93[,4]]
> AVEs  <- (As+Bs+Cs+Ds)/4
```

All squared correlations between the item scores and the average can be calculated at once by binding them together into a matrix and using the correlation function `cor()`:

Correlation function cor()

```
> cor(cbind(As,Bs,Cs,Ds,AVEs))^2
              As          Bs         Cs           Ds       AVEs
As   1.000000000 0.139602528 0.12695057 0.005908244 0.5100255
Bs   0.139602528 1.000000000 0.18681032 0.004365286 0.5793057
Cs   0.126950572 0.186810319 1.00000000 0.047979010 0.6273273
Ds   0.005908244 0.004365286 0.04797901 1.000000000 0.1128582
AVEs 0.510025458 0.579305679 0.62732732 0.112858161 1.0000000
```

The squared correlations (or discrimination measures in homogeneity analysis) on page 157 are recovered in the last column (or row). Their average gives the first principal inertia of the indicator matrix:

```
> sum(cor(cbind(As,Bs,Cs,Ds,AVEs))[1:4,5]^2) / 4
[1] 0.4573792
```

```
> wg93.mca$sv[1]^2
[1] 0.4573792
```

Another result, not mentioned in Chapter 20, is that MCA also maximizes the average covariance between all four item scores. First, calculate the 4×4 covariance matrix between the scores (multiplying by $(N-1)/N$ to obtain the "biased" covariances, since the function `cov()` computes the usual "unbiased" estimates by dividing by $N - 1$), and then calculate the average value of the 16 values using the function `mean()`:

Covariance function cov()

```
> cov(cbind(As,Bs,Cs,Ds)) * 870 / 871
        As          Bs         Cs          Ds
As 1.11510429  0.44403796  0.4406401  0.04031951
Bs 0.44403796  1.26657648  0.5696722  0.03693604
Cs 0.44064007  0.56967224  1.3715695  0.12742741
Ds 0.04031951  0.03693604  0.1274274  0.24674968
```

```
> mean(cov(cbind(As,Bs,Cs,Ds)) * 870 / 871)
[1] 0.4573792
```

Notice that the sum of the variances of the four item scores is equal to 4:

```
> sum(diag(cov(cbind(As,Bs,Cs,Ds)) * 870 / 871))
[1] 4
```

The individual respondents' variance measure in (20.2) is calculated and averaged over the whole sample:

```
> VARs <- ((As-AVEs)^2 + (Bs-AVEs)^2 +( Cs-AVEs)^2 +
+           (Ds-AVEs)^2)/4
> mean(VARs)
[1] 0.5426208
```

which is the loss of homogeneity, equal to 1 minus the first principal inertia.

Exhibit 20.3 can be obtained as the `"rowprincipal"` map, suppressing the row labels (check the plotting options by typing `help(plot.ca)`):

```
> plot(wg93.mca, map="rowprincipal", labels=c(0,2))
```

Chapter 21: Subset Correspondence Analysis

Subset CA in Chapter 21 is presently implemented only in the `ca()` function, but since the Burt matrix is accessible from `mjca()` one can easily do the subset MCA on the Burt matrix. First, an example of subset CA using the `author` dataset, provided with the **ca** package. To reproduce the subset analyses of consonants and vowels:

```
> data(author)
> vowels <- c(1,5,9,15,21)
> consonants <- c(1:26)[-vowels]
> summary(ca(author,subsetcol=consonants))
```

Principal inertias (eigenvalues):

```
dim     value     %     cum%   scree plot
1       0.007607  46.5  46.5   *************************
2       0.003253  19.9  66.4   ***********
3       0.001499   9.2  75.6   *****
4       0.001234   7.5  83.1   ****
.         .        .     .      .
.         .        .     .      .
        -------  -----
Total: 0.01637  100.0
```

Rows:

```
       name   mass  qlt  inr    k=1  cor  ctr    k=2  cor  ctr
1  |   td( |    85   59   29 |    7    8    1 |   -17   50    7 |
2  |   d() |    80  360   37 |  -39  196   16 |   -35  164   31 |
3  |   lw( |    85  641   81 | -100  637  111 |     8    4    2 |
4  |   ew( |    89  328   61 |   17   27    4 |    58  300   92 |
.  |    .  |     .    .    . |    .    .    . |     .    .    . |
.  |    .  |     .    .    . |    .    .    . |     .    .    . |
```

Columns:

```
       name  mass  qlt  inr     k=1  cor  ctr    k=2  cor  ctr
1  |    b |    16   342   21 |   -86  341   15 |    -6    2    0 |
2  |    c |    23   888   69 |  -186  699  104 |   -97  189   66 |
3  |    d |    46   892  101 |   168  783  171 |   -63  110   56 |
4  |    f |    19   558   33 |  -113  467   33 |   -50   91   15 |
.  |    . |     .     .    . |     .    .    . |     .    .    . |
.  |    . |     .     .    . |     .    .    . |     .    .    . |
```

> summary(ca(author, subsetcol=vowels))

Principal inertias (eigenvalues):

```
dim     value     %     cum%   scree plot
1       0.001450  61.4  61.4   *************************
2       0.000422  17.9  79.2   ******
3       0.000300  12.7  91.9   ****
4       0.000103   4.4  96.3   *
5       0.000088   3.7 100.0
        -------  -----
Total: 0.002364 100.0
```

Rows:

```
       name   mass  qlt  inr     k=1  cor  ctr    k=2  cor  ctr
1  |   td( |    85  832  147 |   58  816  195 |    8   15   13 |
2  |   d() |    80  197   44 |  -12  118    9 |  -10   79   20 |
3  |   lw( |    85  235   33 |   14  226   12 |   -3    9    2 |
4  |   ew( |    89  964  109 |   31  337   60 |   42  627  382 |
.  |    .  |     .    .    . |    .    .    . |    .    .    . |
.  |    .  |     .    .    . |    .    .    . |    .    .    . |
```

Columns:

	name	mass	qlt	inr		k=1	cor	ctr		k=2	cor	ctr	
1 |	a |	80	571	79 |		9	34	4 |		-35	537	238 |
2 |	e |	127	898	269 |		67	895	393 |		4	3	5 |
3 |	i |	70	800	221 |		-59	468	169 |		50	332	410 |
4 |	o |	77	812	251 |		-79	803	329 |		-8	9	12 |
5 |	u |	30	694	179 |		-71	359	105 |		-69	334	335 |

We now demonstrate the subset MCA that is documented on pages 165–166, i.e., for the Burt matrix of the working women data set stored in womenG, after elimination of missing data for the demographics (see page 237). First, the function mjca() is used merely to obtain the Burt matrix, and then the subset CA is applied to that square part of the Burt matrix not corresponding to the missing data categories (see re-arranged Burt matrix in Exhibit 21.3). The selection is performed by defining a vector of indices named subset below:

```
> womenG.B <- mjca(womenG)$Burt
> subset <- c(1:16)[-c(4,8,12,16)]
> summary(ca(womenG.B[1:16,1:16], subsetrow=subset,
+                subsetcol=subset))
```

Principal inertias (eigenvalues):

```
dim    value       %    cum%  scree plot
 1     0.263487  41.4   41.4  *************************
 2     0.133342  21.0   62.4  *************
 3     0.094414  14.9   77.3  *********
 4     0.047403   7.5   84.7  *****
 5     0.032144   5.1   89.8  ***
 6     0.026895   4.2   94.0  ***
 7     0.019504   3.1   97.1  **
 8     0.013096   2.1   99.1  *
 9     0.005130   0.8   99.9
10     0.000231   0.0  100.0
11     0.000129   0.0  100.0
       --------  -----
Total: 0.635808  100.0
```

Rows:

	name	mass	qlt	inr		k=1	cor	ctr		k=2	cor	ctr	
1	| Q1.1 |	183	592	25 |		-228	591	36 |		11	1	0 |
2	| Q1.2 |	35	434	98 |		784	345	81 |		-397	88	41 |
3	| Q1.3 |	6	700	119 |		2002	306	88 |		2273	394	224 |
4	| Q2.1 |	13	535	113 |		-1133	236	65 |		1276	299	162 |
5	| Q2.2 |	95	452	69 |		-442	421	71 |		-119	30	10 |
6	| Q2.3 |	120	693	64 |		482	688	106 |		-40	5	1 |
7	| Q3.1 |	28	706	114 |		-1040	412	114 |		878	294	160 |
8	| Q3.2 |	152	481	38 |		-120	91	8 |		-249	390	71 |
9	| Q3.3 |	47	748	106 |		990	681	175 |		312	67	34 |
10	| Q4.1 |	143	731	49 |		-390	702	83 |		80	29	7 |
11	| Q4.2 |	66	583	84 |		582	414	84 |		-371	168	68 |
12	| Q4.3 |	7	702	119 |		1824	312	90 |		2041	391	222 |

The adjustments of the scale by linear regression to best fit the off-diagonal tables of the Burt submatrix is not easily done. The code is rather lengthy, so is not described here but rather put on the website for the moment, to be implemented in the **ca** package at a later date.

As shown in Chapter 21 the CA of a square asymmetric matrix consists in splitting the table into symmetric and skew-symmetric parts and then performing CA on the symmetric part and an uncentred CA on the skew-symmetric part, with the same weights and χ^2-distances throughout. Both analyses are neatly subsumed in the CA of the block matrix shown in (22.4). After reading the mobility table into a data frame named `mob`, the sequence of commands to set up the block matrix and then do the CA is as follows. Notice that `mob` has to be first converted to a matrix; otherwise we cannot bind the rows and columns together properly to create the block matrix `mob2`.

Chapter 22:
Analysis of Square
Tables

```
> mob <- as.matrix(mob)
> mob2 <- rbind(cbind(mob,t(mob)), cbind(t(mob), mob))
> summary(ca(mob2))
```

Principal inertias (eigenvalues):

dim	value	%	cum%	scree plot
1	0.388679	24.3	24.3	************************
2	0.232042	14.5	38.8	***************
3	0.158364	9.9	48.7	**********
4	0.158364	9.9	58.6	**********
5	0.143915	9.0	67.6	*********
6	0.123757	7.7	75.4	********
7	0.081838	5.1	80.5	*****
8	0.070740	4.4	84.9	*****
9	0.049838	3.1	88.0	***
10	0.041841	2.6	90.6	***
11	0.041841	2.6	93.3	***
12	0.022867	1.4	94.7	*
13	0.022045	1.4	96.1	*
14	0.012873	0.8	96.9	*
15	0.012873	0.8	97.7	*
16	0.010360	0.6	98.3	*
17	0.007590	0.5	98.8	*
18	0.007590	0.5	99.3	*
19	0.003090	0.2	99.5	
20	0.003090	0.2	99.7	
21	0.001658	0.1	99.8	
22	0.001148	0.1	99.9	
23	0.001148	0.1	99.9	
24	0.000620	0.0	99.9	
25	0.000381	0.0	100.0	
26	0.000381	0.0	100.0	
27	0.000147	0.0	100.0	

```
         -------- -----
Total: 1.599080 100.0
```

Rows:

	name	mass	qlt	inr	k=1	cor	ctr	k=2	cor	ctr	
1	Arm	43	426	54	-632	200	44	671	226	84	
2	Art	55	886	100	1521	793	327	520	93	64	
3	Tcc	29	83	10	-195	73	3	73	10	1	
4	Cra	18	293	32	867	262	34	-298	31	7	
.	.	.	.	.	.	.	.	.	.	.	
.	.	.	.	.	.	.	.	.	.	.	
15	ARM	43	426	54	-632	200	44	671	226	84	
16	ART	55	886	100	1521	793	327	520	93	64	
17	TCC	29	83	10	-195	73	3	73	10	1	
18	CRA	18	293	32	867	262	34	-298	31	7	
.	.	.	.	.	.	.	.	.	.	.	
.	.	.	.	.	.	.	.	.	.	.	

Columns:

	name	mass	qlt	inr	k=1	cor	ctr	k=2	cor	ctr	
1	ARM	43	426	54	-632	200	44	671	226	84	
2	ART	55	886	100	1521	793	327	520	93	64	
3	TCC	29	83	10	-195	73	3	73	10	1	
4	CRA	18	293	32	867	262	34	-298	31	7	
.	.	.	.	.	.	.	.	.	.	.	
.	.	.	.	.	.	.	.	.	.	.	
15	Arm	43	426	54	-632	200	44	671	226	84	
16	Art	55	886	100	1521	793	327	520	93	64	
17	Tcc	29	83	10	-195	73	3	73	10	1	
18	Cra	18	293	32	867	262	34	-298	31	7	
.	.	.	.	.	.	.	.	.	.	.	
.	.	.	.	.	.	.	.	.	.	.	

The principal inertias coincide with Exhibit 22.4, and since the first two dimensions correspond to the symmetric part of the matrix, each set of coordinates is just a repeat of the same set of values.

Dimensions 3 and 4, with repeated eigenvalues, correspond to the skew-symmetric part and their coordinates turn out as follows (to get more than the default two dimensions in the summary, change the original command to `summary(ca(mob2, nd=4))`):

Rows:

	name	k=3	cor	ctr	k=4	cor	ctr	
1	Arm	-11	0	0	416	87	47	
2	Art	89	3	3	423	61	62	
3	Tcc	-331	211	20	141	38	4	
4	Cra	-847	250	80	92	3	1	
.	.	.	.	.	.	.	.	
.	.	.	.	.	.	.	.	
15	ARM	11	0	0	-416	87	47	
16	ART	-89	3	3	-423	61	62	
17	TCC	331	211	20	-141	38	4	
18	CRA	847	250	80	-92	3	1	

```
. .   . .   . .
.   . .   . .   .
```

```
Columns:
       name     k=3 cor ctr     k=4 cor ctr
 1 |  ARM |    -416  87  47 |    -11   0   0 |
 2 |  ART |    -423  61  62 |     89   3   3 |
 3 |  TCC |    -141  38   4 |   -331 211  20 |
 4 |  CRA |     -92   3   1 |   -847 250  80 |
 .    .      .   .   .      .   .   .
 .    .      .   .   .      .   .   .

15 |  Arm |     416  87  47 |     11   0   0 |
16 |  Art |     423  61  62 |    -89   3   3 |
17 |  Tcc |     141  38   4 |    331 211  20 |
18 |  Cra |      92   3   1 |    847 250  80 |
 .    .      .   .   .      .   .   .
 .    .      .   .   .      .   .   .
```

which shows that the skew-symmetric coordinates reverse sign within the row
and column blocks, but also swap over, with the third axis row solution equal
to the fourth axis column solution and vice versa. In any case, only one set of
coordinates is needed to plot the objects in each map, but the interpretation
of the maps is different, as explained in Chapter 22.

Chapter 23: Data Recoding

Chapter 23 involves mostly simple transformations of the data and then regular applications of CA. As an illustration of analyzing continuous data, we
assume that the European Union indicators data have been read into a data
frame named EU. Then the conversion to ranks (using R function `rank()` and
again the very useful `apply()` function to obtain EUr), and the doubling (to
obtain EUd) are performed as follows:

Converting to ranks with `rank()` function

```
> EUr <- apply(EU,2,rank) - 1
> EUd <- cbind(EUr, 11-EUr)
> colnames(EUd) <- c(paste(colnames(EU), "-", sep=""),
+                     paste(colnames(EU), "+", sep=""))
> EUd
```

	Unemp-	GDPH-	PCH-	PCP-	RULC-	Unemp+	GDPH+	PCH+	PCP+	RULC+
Be	6	6	6	6.5	4.5	5	5	5	4.5	6.5
De	4	11	10	0.0	7.0	7	0	1	11.0	4.0
Ge	2	10	11	5.0	6.0	9	1	0	6.0	5.0
Gr	5	1	1	1.0	11.0	6	10	10	10.0	0.0
Sp	11	3	3	10.0	2.0	0	8	8	1.0	9.0
Fr	7	8	8	3.5	4.5	4	3	3	7.5	6.5
Ir	10	2	2	11.0	1.0	1	9	9	0.0	10.0
It	9	7	7	9.0	9.0	2	4	4	2.0	2.0
Lu	0	9	9	3.5	8.0	11	2	2	7.5	3.0
Ho	8	5	4	6.5	3.0	3	6	7	4.5	8.0
Po	1	0	0	8.0	0.0	10	11	11	3.0	11.0
UK	3	4	5	2.0	10.0	8	7	6	9.0	1.0

Notice how the column names are constructed with the `paste()` function.
The analysis of Exhibit 23.5 is thus obtained simply as `ca(EUd)`.

The results of Chapter 24 cannot be obtained using the **ca** package, but using either the XLSTAT program (described later) or Jari Oksanen's **vegan** package (see web resources in the Bibliographical Appendix), which not only does CCA but also CA and PCA (but without many of the options we have in the **ca** package). Since this package is usually used in an ecological context, like the example in Chapter 24, we shall refer here to "sites" (samples), "species" and (explanatory) "variables". Using **vegan** is just as easy as using **ca**: the main function is called `cca()` and can be used in either of the two following formats:

```
cca(X,Y,Z)
cca(X ~ Y + condition(Z))
```

where X is the sites×species matrix of counts, Y is the sites×variables matrix of explanatory data and Z is the sites×variables matrix of conditioning data if we want to perform (optionally) a partial CCA. The second format is in the form of a regression-type model formula, but here the first type will be used. If only X is specified, the analysis is a CA (so try, for example, one of the previous analyses, for example `summary(cca(author))` to compare the results with previous ones — notice that the books are referred to as "sites" and the letters as "species", and that the default plotting option, for example `plot(cca(author))`, is what we called `"colprincipal"`). If X and Y are specified, the analysis is a CCA. If X, Y and Z are specified, the analysis is a partial CCA.

Assuming now that the biological data of Chapters 10 and 24 are read into the data frame `bio` as a 13×92 table, and that the three variables Ba, Fe and PE are read into `env` as a 13 × 3 table whose columns are log-transformed to variables with names `logBa`, `logFe` and `logPE`; then the CCA can be performed simply as follows:

```
> summary(cca(bio, env))

Call:
cca(X = bio, Y = env)

Partitioning of mean squared contingency coefficient:

Total         0.7826
Constrained   0.2798
Unconstrained 0.5028

Eigenvalues, and their contribution to the
            mean squared contingency coefficient

             CCA1    CCA2    CCA3    CA1     CA2     CA3     CA4     CA5
lambda      0.1895  0.0615  0.02879 0.1909  0.1523  0.04159 0.02784 0.02535
accounted   0.2422  0.3208  0.35755 0.2439  0.4385  0.49161 0.52719 0.55957
             CA6     CA7     CA8     CA9
lambda      0.02296 0.01654 0.01461 0.01076
accounted   0.58891 0.61004 0.62871 0.64245

Scaling 2 for species and site scores
--- Species are scaled proportional to eigenvalues
--- Sites are unscaled: weighted dispersion equal on all dimensions
```

Species scores

	CCA1	CCA2	CCA3	CA1	CA2	CA3
Myri_ocul	0.1732392	0.245915	-0.070907	0.6359626	-0.063479	0.031990
Chae_seto	0.5747974	-0.270816	0.011814	-0.5029157	-0.674207	0.093354
Amph_falc	0.2953878	-0.114067	0.075979	-0.2224138	0.041797	-0.005020
Myse_bide	-0.5271092	-0.505262	-0.103978	-0.0789909	0.176683	-0.484208
Goni_macu	-0.1890403	0.122783	-0.044679	-0.1045244	0.030134	0.111827
Amph_fili	-0.9989672	-0.075696	0.107184	-0.3506103	0.076968	0.004931
.	.	.	.	.	.	.
.	.	.	.	.	.	.

Site constraints (linear combinations of constraining variables)

	CCA1	CCA2	CCA3
S4	-0.06973	0.75885	-2.29951
S8	-0.35758	1.47282	2.27467
S9	0.48483	-0.72459	-0.66547
S12	0.02536	0.27129	-0.14677
S13	0.30041	-0.01531	-0.80821
S14	0.79386	1.16229	0.24314
S15	0.96326	-0.88970	0.14630
S18	-0.16753	0.25048	-0.77451
S19	0.36890	-0.81800	1.50620
S23	-0.09967	-1.90159	0.06877
S24	0.05478	0.96184	-0.10635
R40	-3.71393	-0.20698	0.53031
R42	-2.96641	-0.18264	-0.67736

Biplot scores for constraining variables

	CCA1	CCA2	CCA3
logBa	0.9957	-0.08413	0.03452
logFe	0.6044	-0.72088	0.33658
logPE	0.4654	0.55594	0.68710

Notice the following:

— the `mean squared contingency coefficient` is the total inertia;

— the principal inertias in the constrained space are headed CCA1, CCA2, etc., and the principal inertias in the unconstrained space CA1, CA2, etc.;

— the percentages are all expressed relative to the total inertia;

— Scaling 2 means rows (sites) in standard coordinates, and columns (species) in principal coordinates, i.e., the "colprincipal" scaling in the plot.ca() function;

— the Species scores are column principal coordinates;

— the Site constraints are the row standard coordinates;

— the Biplot scores for constraining variables are the weighted correlation coefficients between the explanatory variables and the site coordinates.

Chapter 25:
Aspects of
Stability and
Inference

Finally, Chapter 25 performs various bootstraps of tables to investigate their
variability, as well as permutation tests to test null hypotheses. For example,
the 1000 replications of the author data, shown in the partial bootstrap CA
map of Exhibit 25.1, are obtained as follows (with comments interspersed):

```
> data(author)
> author.ca <- ca(author)
> nsim <- 1000
> # compute row sums
> author.rowsum <- apply(author, 1, sum)
> # compute nsim simulations of first book
> author.sim <- rmultinom(nsim, author.rowsum[1], prob = author[1,]
```

Multinomial
random sampling
using
rmultinom()

```
> # compute nsim simulations of other books and column-bind
> for (i in 2:12) {
+   author.sim<-cbind(author.sim,
+                     rmultinom(nsim,author.rowsum[i],
+                         prob = author[i,]))
+ }
> # transpose to have same format as original matrix
> author.sim <- t(author.sim)
> author.sim2 <- matrix(rep(0,nsim*12*26), nrow=nsim*12)
> # reorganize rows so that matrices are together
> for (k in 1:nsim) {
+   for (i in 1:12) {
+     author.sim2[(k-1)*12+i,] <- author.sim[k+(i-1)*nsim,]
+     }
+ }
```

The coordinates are now calculated of the simulated columns using the transition formula from row standard to column principal coordinates:

```
> # get standard coordinates for rows
> author.rowsc <- author.ca$rowcoord[,1:2]
> # calculate pc's of all replicates using transition formula
> author.colsim <- t(t(author.rowsc) %*% author.sim2[1:12,]) /
+                   apply(author.sim2[1:12,],2,sum)
> for (k in 2:nsim) {
+   author.colsim <- rbind(author.colsim, t(t(author.rowsc) %*%
+                       author.sim2[((k-1)*12+1):(k*12),])/
+                       apply(author.sim2[((k-1)*12+1):(k*12),], 2, sum
> # reorganize rows of coordinates so that letters are together
> author.colsim2 <- matrix(rep(0, nsim*26*2), nrow=nsim*26)
> for (j in 1:26) {
+   for (k in 1:nsim) {
+     author.colsim2[(j-1)*nsim+k,]<-author.colsim[j+(k-1)*26,]
+     }
+ }
```

The plotting of the points and the convex hulls:

```
> # plot all points (use first format of coords for labelling...)
> plot(author.colsim[,1], -author.colsim[,2], xlab="dim1",
+      ylab="dim2",type="n")
> text(author.colsim[,1], -author.colsim[,2], letters, cex=0.5, col="gray")
> # plot convex hulls for each letter
> # first calculate pc's of letters for original matrix
> author.col <- t(t(author.rowsc) %*% author) /
+                  apply(author, 2, sum)
> for (j in 1:26) {
+  points <- author.colsim2[(100*(j-1)+1):(100*j),]
+ # note we are reversing second coordinate in all these plots
+  points[,2] <- -points[,2]
+  hpts <- chull(points)
+  hpts <- c(hpts,hpts[1])
+  lines(points[hpts,], lty=3)
+  text(author.col[j,1], -author.col[j,2], letters[j], font=2, cex=1.5)
+ }
```

Finally, peeling away the convex hulls until just under 5% of the points have been removed from each cloud of replicates, and then plotting the convex hulls of the remaining points:

```
> plot(author.colsim2[,1], -author.colsim2[,2], xlab="dim1",
+      ylab="dim2", type="n")
> for (j in 1:26) {
+  points <- author.colsim2[(100*(j-1)+1):(100*j),]
+ # note we are reversing second coordinate in all these plots
+  points[,2] <- -points[,2]
+  repeat {
+      hpts <- chull(points)
+      npts <- nrow(points[-hpts,])
+      if(npts/nsim<0.95) break
+      points <- points[-hpts,]
+         }
+  hpts <- c(hpts,hpts[1])
+  lines(points[hpts,], lty=3)
+  text(author.col[j,1], -author.col[j,2], letters[j],
+       font=2)
+ }
```

To plot concentration ellipses, the package **ellipse** needs to be downloaded from the R website www.R-project.org. The script for plotting concentration ellipses using the bootstrap replicates computed above is as follows:

```
> # confidence ellipses - needs package 'ellipse'
> plot(author.colsim2[,1],-author.colsim2[,2],xlab="dim1",
+      ylab="dim2", type="n")
> for (j in 1:26) {
```

```
+    points <- author.colsim2[(nsim*(j-1)+1):(nsim*j),]
+    # note we are reversing second coordinate in all these plots
+    points[,2] <- -points[,2]
+    covpoints <- cov(points)
+    meanpoints <- apply(points,2,mean)
+    lines(ellipse(covpoints, centre = meanpoints))
+    text(author.col[j,1],-author.col[j,2],letters[j],
+        font=2)
+                    }
```

To reproduce the ellipses based on the Delta method in Exhibit 25.3, the covariance matrix of the estimated principal coordinates is needed, which is provided in the output of the SPSS program. More details and additional R code are given on the CARME network website www.carme-n.org.

Permutation test The permutation test on the author data is performed by exhaustively listing the $11 \times 9 \times 7 \times 5 \times 3 = 10395$ different ways of assigning the labels of the texts to the points in the CA map, recalculating the sum of distance between pairs of labels by the same author. The R script for obtaining every unique assignment is given on the website: it consists of listing the assignments $\{(1,2),(1,3),(1,4),...,(1,12)\}$ and then for each of these 11, listing 9 other pairings (e.g., for (1,2) the list would be $\{(3,4),(3,5),...,(3,12)\}$), and then for each of these 9 a similar list of 7 formed by the remaining integers, and so on. Exhibit B.6 shows the distribution of the 10395 values, with the observed value of 0.4711 indicated. As reported in Chapter 25, there are no other pairings of the labels in the two-dimensional CA map which give a smaller distance, which makes the *P*-value associated with the result equal to 1/10395, i.e., $P < 0.0001$. A similar test conducted on the subset analyses of Exhibit 21.1 (consonants only) and 21.2 (vowels only) yielded 47 and 4533 permutations to the left of the observed value, i.e., *P*-values of $47/10395 = 0.0045$ and $4533/10395 = 0.44$, respectively. This shows that the consonants are the ones that are distinguishing the authors, not the vowels.

Permutation tests in CCA CCA looks at a part of the full space of the response variables (usually species abundances in ecology) that is linearly related to a given set of explanatory variables (usually environmental variables) — see Chapter 24. But what if the responses have no relation to the explanatory variables? The inertia in the constrained space is a measure of the relationship, so this needs to be compared to a null distribution of constrained inertias when there is no relationship. This can be generated by randomly permuting the cases (rows) in the matrix of explanatory variables (or response variables). Having mixed up the rows, losing their connections with the rows in the response matrix, the CCA is repeated and the constrained inertia recalculated. Doing this 999 times, say (or however many you need to establish an accurate *P*-value), the observed value is situated in the distribution to see if it is unusually high. If it lies in the top 5% of observations, the relationship will be deemed statistically significant, and the *P*-value can be estimated by counting how many values in the permutation distribution are higher than the observed value (in this case

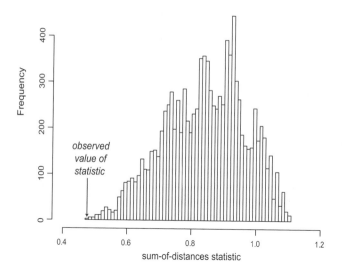

Exhibit B.6:
Exact null distribution of sum-of-distances statistic in the permutation test for testing randomness in the positions of pairs of texts by the same author in the CA map. The observed value is the second smallest out of 10395 possible values.

the value has to be high to be significant). The **vegan** package incorporates this test, which can be obtained with the `anova()` function around the `cca()` analysis:

```
> anova(cca(bio,env))

Permutation test for cca under reduced model

Model: cca(X = bio, Y = env)
        Df  Chisq      F N.Perm  Pr(>F)
Model    3 0.2798 1.6696    1300 0.03462 *
Residual 9 0.5028
---
Signif. codes:  0 '***' 0.001 '**' 0.01 '*' 0.05 '.' 0.1 ' ' 1
```

The statistic used is not the inertia but a "pseudo-F" statistic, as if an analysis of variance is being performed (for more details, see the **vegan** documentation), but the important part of the printout is the *P*-value which shows that the F-statistic for the constrained space is significantly high (P=0.03462).

In his recent book *Correspondence Analysis and Data Coding with Java and R* (see Bibliographical Appendix), Fionn Murtagh gives many R scripts for CA and especially data recoding, all of which are available on the website `www.correspondances.info`. In particular, on pages 21–26 he describes a program for hierarchical clustering by Ward's method, with incorporation of weights, which is exactly what we needed in Chapter 15, but which is otherwise unavailable in R. Assuming you have been able to download the code from his website, and have read the table of data of Exhibit 15.3 as the data frame `food`, then the cluster analysis of the row profiles in Exhibit 15.5 can be achieved using Murtagh's `hierclust()` function as follows:

```
> food.rpro <- food /apply(food,1,sum)
> food.r <- apply(food,1,sum) / sum(food)
> food.rclust <- hierclust(food.rpro, food.r)
> plot(as.dendrogram(food.rclust))
```

XLSTAT   One of the best commercial alternatives for performing all the analyses in this book, and more, is the suite of Excel add-on packages called XLSTAT (www.xlstat.com). The CA and MCA programs in XLSTAT now include the adjustments of inertia in MCA and the subset options in both CA and MCA. A program for CCA is also implemented, including a permutation test for testing whether the explanatory variables are significantly related to the principal axes of the constrained solution. Other multivariate analysis software in XLSTAT includes principal component analysis, factor analysis, discriminant analysis, cluster analysis, partial least squares and generalized Procrustes analysis. The programs are particularly easy to use because they function in an Excel environment. For example, to execute the CA on the food data set used above in the hierarchical clustering, click on the CA icon and select the table (with row and column labels) that you want to analyze:

Exhibit B.7:
XLSTAT menu for executing CA on a table selected in Excel.

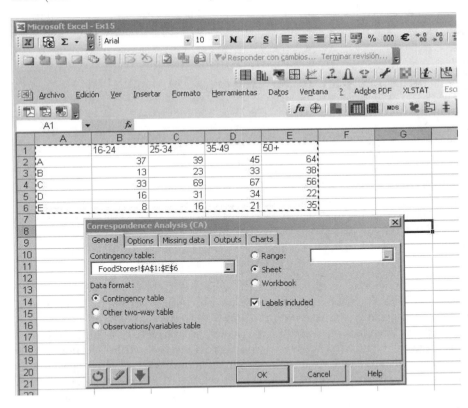

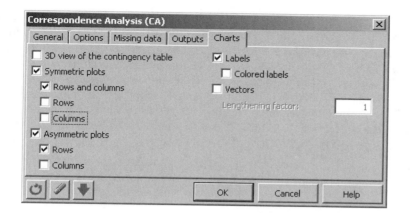

Exhibit B.8:
XLSTAT menu for selecting graphical options in CA.

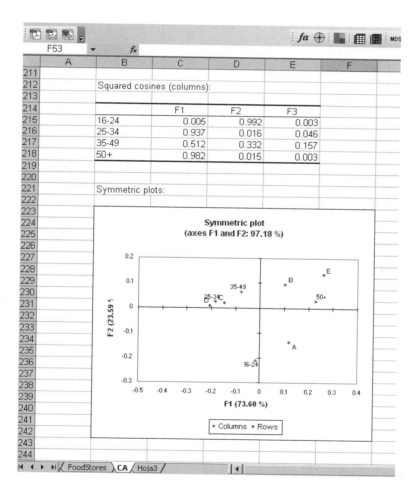

Exhibit B.9:
Part of the output of the CA program in XLSTAT, which is returned in a separate worksheet.

There is an "Options" menu which allows selection of supplementary points or subsets, a "Missing data" menu for deciding what to do with missing values, and "Outputs" menu for selecting the various numerical tables (profiles, χ^2-distances, principal coordinates, standard coordinates, contributions, squared correlations, etc.) and a "Charts" menu which allows selection of the various maps — see Exhibit B.8, where the symmetric map of rows and columns is selected as well as the asymmetric map of the rows (i.e., `"rowprincipal"` option in the **ca** package). Exhibit B.9 shows part of the output.

In the cluster analysis module of XLSTAT it is possible to assign weights to points, so performing a Ward clustering on the profiles, weighted by their masses, would reproduce the cluster analyses of Chapter 15.

Graphical options   Producing a CA map with certain characteristics and which is ready for publication is not a trivial task. In this section we describe the three different technologies that were used in this book to produce the graphical exhibits.

LaTeX graphics   This book was typeset in LaTeX. LaTeX itself and various LaTeX macros can produce maps directly, without passing through another graphics package. Most of the maps produced in the first part of the book were produced using the macro package PicTeX. As an example, the following code, which is embedded in the LaTeX text of the book itself, produced Exhibit 9.2, the asymmetric map of the smoking data:

```
\beginpicture
\setcoordinatesystem units <2.5cm,2.5cm>
\setplotarea x from -2.40 to 1.70, y from -1.6 to 2.25
\accountingoff
\gray
\setdashes <5pt,4pt>
\putrule from 0 0 to 1.7 0
\putrule from 0 0 to -1.4 0
\putrule from 0 0 to 0 2.25
\putrule from 0 0 to 0 -1.6
\put {+} at 0 0
\black
\small
\put {Axis 1} [Br] <-.2cm,.15cm> at 1.70 0
\put {0.0748 (87.8\%)} [tr] <-.2cm,-.15cm> at 1.70 0
\put {Axis 2} [Br] <-.1cm,-.4cm> at 0 2.25
\put {0.0100 (11.8\%)} [Bl] <.1cm,-.4cm> at 0 2.25
\setsolid
\putrule from 1.3  -1.3  to 1.4  -1.3
\putrule from 1.3  -1.32 to 1.3  -1.28
\putrule from 1.4  -1.32 to 1.4  -1.28
\put {\it scale} [b] <0cm,.25cm> at 1.35 -1.3
\put {0.1} [t] <0cm,-.2cm> at   1.35 -1.3
\multiput {$\bullet$} at
 0.06577   0.19373
-0.25896   0.24330
```

```
  0.38059   0.01066
 -0.23295  -0.05775
  0.20109  -0.07891
  /
\sf
\put {SM} [l] <.15cm,0cm> at   0.06577   0.19373
\put {JM} [r] <-.15cm,0cm> at  -0.25896   0.24330
\put {SE} [bl] <.15cm,0cm> at   0.38059   0.01066
\put {JE} [r] <-.15cm,0cm> at  -0.23295  -0.05775
\put {SC} [tl] <.15cm,0cm> at   0.20109  -0.07891
\gray
\multiput {$\circ$} at
  1.4384    0.3046
 -0.3638   -1.4094
 -0.7180   -0.0735
 -1.0745    1.9760
  /
\sl
\put {none} [b] <0cm,.2cm> at      1.4384    0.3046
\put {light} [b] <0cm,.2cm> at    -0.3638   -1.4094
\put {medium} [T] <0cm,-.3cm> at  -0.7180   -0.0735
\put {heavy} [b] <0cm,.2cm> at    -1.0745    1.9760
\black
\endpicture
```

Comparing the above code with Exhibit 9.2 itself should be enough for you to see how each line and each character is laboriously placed in the plotting area. One advantage of this approach, however, is that once you have set the units on the horizontal and vertical coordinate axes to be the same (2.5cm per unit in the example above), then you are assured that the aspect ratio of 1 is perfectly preserved in the eventual result.

Excel graphics

Since many of the new maps produced for this second edition were made in Excel using XLSTAT, many exhibits are produced in this style, for example those in Chapters 17–19. In Excel, a certain amount of trimming of the maps was done, redefining the maxima and minima on the axes, and also stretching the graph window vertically or horizontally until the aspect ratio appeared correct. The graphic was then copied as a metafile and pasted into *Adobe Illustrator*, where further trimming and character redefinition was performed. The aspect ratio becomes slightly deformed when copying into *Adobe Illustrator*, with a vertical unit appearing slightly longer than a horizontal unit, so some resizing is necessary at this stage as well. The graphic was then saved as an *Encapsulated PostScript* (EPS) file and then included in the LaTeX file using the \includegraphics instruction, for example:

```
\begin{figure}[h]
\center{\includegraphics[width=10cm,keepaspectratio]{Ex18_5.eps}}
\caption{\sl MCA map of Burt matrix of four questions on women
        working, showing first and second dimensions;
        total inertia = 1.145, percentage inertia in map: 65.0\%.}
\end{figure}
```

R *graphics*   Finally, many maps were also produced in R, for example all the graphics of Chapter 25. These were also copied as metafiles and pasted into *Adobe Illustrator*, fine-tuned and then saved as an EPS file for including in the text.

Bibliography of Correspondence Analysis

The main aim of this book is to teach the reader about CA. No references have been given in the text since this detracts from the didactic purpose. This section is meant to highlight the main bibliographical sources so that the reader can continue to learn about this method. We shall also point out where historical details about the method can be obtained as well as more complete literature reviews.

Although the theory of CA dates back to much earlier in the 20th century, CA as presented in this book originates in the work of Jean-Paul Benzécri and his co-workers in France in the 1960s, which was published in the two volumes of *Analyse des Données* (literally, Data Analysis):

Benzécri's School of Data Analysis

—Benzécri, J.-P. & collaborateurs (1973) *Analyse des Données. Tôme 1: La Classification. Tôme 2: L'Analyse des Correspondances.* Paris: Dunod.

However, these books remain inaccessible to readers who are not initiated into Benzécri's particular notational style, preferred over more pragmatic matrix-vector notation. An English translation of these books has appeared, but with little success in communicating Benzécri's ideas to an English-speaking community. The following book by Le Roux & Rouanet gives a good account of Benzécri's approach for analyzing large data sets, which the authors have coined as "geometric data analysis", although they have also maintained a complex notational style that hinders understanding of the material:

—Le Roux, B. & Rouanet, H. (2004) *Geometric Data Analysis: From Correspondence Analysis to Structured Data.* Dordrecht: Kluwer.

One of the best publications in English for understanding Benzécri's work is the book by Fionn Murtagh, who was also a student of Benzécri. Not only does he communicate much more of the Benzecrian philosophy (there is also a foreword by Benzécri himself, with an English translation), but also the book is innovative in approach and highly computing-oriented, providing many interesting applications and details of R programming.

—Murtagh, F. (2005) *Correspondence Analysis and Data Coding with Java and R.* London: Chapman & Hall/CRC.

One of the leading and most innovative members of Benzécri's group, Brigitte Escofier, has been commemorated posthumously by a collection of her most important articles:

—Escofier, B. (2003) *Analyse des Correspondances: Recherches au Coeur de l'Analyse des Donées.* Rennes, France: Presses Universitaires de Rennes.

The two English books In 1984 two English books on CA appeared almost simultaneously, expressing
of 1984 Benzécri's work in a more comprehensible way, thanks to the use of more
conventional mathematical notation:

— Lebart, L., Morineau, A. & Warwick, K. (1984) *Multivariate Descriptive
  Statistical Analysis.* Chichester, UK: Wiley.

— Greenacre, M.J. (1984) *Theory and Applications of Correspondence Analysis.* London: Academic Press.

Both these books are out of print, but are worth consulting if available in a
library. Lebart et al.'s book gives a less detailed description of CA itself but a
broader view of its use in the context of large-scale social surveys. Greenacre's
book attempts to be complete account of the method's theory and practice
at that time. Both these books serve as good literature sources for work up
to that point in time.

The Gifi system Under the *nom-de-plume* of Albert Gifi, a group in Holland, led by Jan de
Leeuw, was involved with the most important development of CA outside
France, and still remains the most active group today. This group mostly
explored the use of MCA — which it called *homogeneity analysis* — as a
quantification technique embedded in classical multivariate analysis to achieve
nonlinear generalizations of multivariate methods. The work of the Gifi group
is amply described in the book:

— Gifi, A. (1990) *Nonlinear Multivariate Analysis.* Chichester, UK: Wiley.

As an excellent summary of the "Gifi system", see:

— Michalidis, G. & de Leeuw, J. (1998) The Gifi system for descriptive multivariate analysis. *Statistical Science*, **13**, 307–336. (This article can be
  googled).

The Japanese school Founded by Chikio Hayashi, this group developed, in parallel to the French
and Dutch schools, an equivalent system of data analysis called "quantification of qualitative data", imbued with its own cultural aspects. Several books
by Shizuhiko Nishisato describe this approach, renamed as "dual scaling",
concentrating more on the algebraic properties of the quantified scale values,
although the recent book by Nishisato does contain many graphical displays:

— Nishisato, S. (2006) *Multivariate Nonlinear Descriptive Analysis.* London:
  Chapman & Hall/CRC.

Nishisato's book contains many historical details and a very comprehensive
reference list of CA-related literature, but no details about computing.

Cologne and Barcelona In 1991, 1995, 1999 (at the Central Archive for Empirical Social Research,
books Cologne) and 2003 (at the Universitat Pompeu Fabra, Barcelona) international conferences with CA as the central theme took place. As a product
of three of these conferences, books were collectively written by statisticians
and social scientists to reflect the development of the theoretical and practical
aspects of CA and related methods:

—Greenacre, M.J. & Blasius, J., editors (1994) *Correspondence Analysis in the Social Sciences*. London: Academic Press.

—Blasius, J. & Greenacre, M.J., editors (1998) *Visualizing Categorical Data*. San Diego: Academic Press.

— Greenacre, M.J. & Blasius, J., editors (2006) *Multiple Correspondence Analysis and Related Methods*. London: Chapman & Hall/CRC.

These three volumes, to which over 100 authors have contributed, are highly recommended for further reading. The third volume is particularly oriented to computing needs as well, and many sources of computer software are given by the individual authors.

The aim of this second edition of *Correspondence Analysis in Practice* is not only to present a didactically structured text about the method, but also to enable readers to compute their own analyses, mostly using the R programming system, which has become the standard of statistical computing at the start of the 21st century. Apart from the additional material included here compared to the first edition, the addition of the 46-page Computational Appendix is the most significant change. One of the many books on R programming which we can recommend to anyone starting off, as well as an excellent introduction to modern statistical methodology, is:

The R connection

—Crawley, M. (2005). *Statistics: An Introduction using R*. Chichester, UK: Wiley.

The **ca** package for R is discussed in more detail in the following article:

—Nenadić, O. & Greenacre, M.J. (2007). Correspondence analysis in R, with two- and three-dimensional graphics: The **ca** package. *Journal of Statistical Software*. Free download from `http://www.jstatsoft.org`.

The following (non-commercial) pages can be consulted for further information and software about CA and related methods:

Web resources

— `http://www.carme-n.org`

   (Correspondence Analysis and Related Methods Network, with R scripts and data from *Correspondence Analysis in Practice, Second Edition*)

— `http://gifi.stat.ucla.edu`

   (Jan de Leeuw's website for the Gifi system and R functions)

— `http://www.correspondances.info`

   (Fionn Murtagh's website for his book, with R scripts and data sets)

— `http://www.math.yorku.ca/SCS/friendly.html`

   (Michael Friendly's personal page for graphics of categorical data)

— `http://www.imperial.ac.uk/bio/research/crawley/statistics`

   (Michael Crawley's material from his book *Statistics: an Introduction using R*)

— `http://www.gesis.org/en/za`

(website of Central Archive for Empirical Social Research in Cologne, with links to various social surveys including the those of the ISSP – International Social Survey Program)

— `http://www.r-project.org`

(The R project for statistical computing)

— `http://cc.oulu.fi/ jarioksa/softhelp/vegan.html`

(Jari Oksanen's website for the **vegan** package in R)

— `http://people.few.eur.nl/groenen/mmds/datasets`

(website with data sets from book *Modern Multidimensional Scaling* by Ingwer Borg and Patrick Groenen)

Glossary of terms

In this appendix an alphabetical list of the most common terms used in this book is given, along with a short definition of each. Words in italics refer to terms which are contained in the glossary.

- *adjusted principal inertias* — a modification of the results of a *multiple correspondence analysis* that gives a more realistic estimate of the inertia accounted for in the solution.

- *arch effect* — the tendency for points in a CA map to form a curve, owing to the particular geometry of CA where the profiles lie inside a *simplex*; also called the "horseshoe" effect.

- *aspect ratio* — the ratio between a unit length on the horizontal axis and a unit length on the vertical axis in a spatial representation; should be equal to 1 for a CA *map*.

- *asymmetric map* — a joint display of the rows and columns where the two clouds of points have different normalizations (also called scalings), usually one in *principal coordinates* and the other in *standard coordinates*; the asymmetric map is often a *biplot*.

- *biplot* — a joint display of points representing the rows and columns of a table such that *scalar products* between a row point and a column point approximates optimally the corresponding element in the table.

- *biplot axis* — a line in the direction of a point vector in a *biplot* onto which the other set of points can be projected in order to estimate values in the table being analysed.

- *bootstrapping* — a computer-based method of investigating the variability of a statistic, by generating a large number replicate samples from the observed sample.

- *Burt matrix* — a particular matrix of *stacked tables*, consisting of all two-way cross-tabulations of a set of Q categorical variables, including the cross-tabulations of each variable with itself.

- *calibration* — in *biplots*, the process of putting a scale on a *biplot axis* with specific tic-marks and values; in CA, where *profiles* are being mapped, this is a scale in units of proportions or percentages.

- *canonical correspondence analysis (CCA)* — extension of CA to include external explanatory variables; the CA solution is constrained to have dimensions which are linearly related to these explanatory variables.

- *centroid* — the weighted average point.

- *chi-square distance* — weighted *Euclidean distance* measure between *profiles*, where each squared difference between profile elements is divided by the corresponding element of the average profile.

- *chi-square statistic* — the statistic used commonly for testing the *independence model* for a *contingency table*; calculated as the sum of squared differences between observed frequencies and frequencies expected according to the model, each squared difference being divided by the corresponding expected frequency.

- *contingency ratio* — for a *contingency table*, the observed frequency divided by the expected frequency according to the *independence model*.

- *contingency table* — a cross-tabulation of a set of individuals according to two categorical variables; hence the grand total of the table is the number of individuals.

- *contribution to inertia* — component of *inertia* accounted for by a particular point on a particular *principal axis*; these are usually expressed relative to the corresponding *principal inertia* on the axis (giving a diagnostic of how the axis is constructed) or relative to the inertia of the point (giving a measure of how well the point is explained by the axis).

- *correspondence analysis (CA)* — a method of displaying the rows and columns of a table as points in a spatial map, with a specific geometric interpretation of the positions of the points as a means of interpreting the similarities and differences between rows, the similarities and differences between columns and the association between rows and columns.

- *dimensionality* — the number of dimensions inherent in a table needed to reproduce its elements exactly in a CA *map*.

- *doubling* — a recoding scheme where a row (or column) is recoded as a pair of rows (or columns) in order to map the extremes, or poles, of a scale; used in CA to analyse ratings, preferences and paired comparisons.

- *dummy variable* — a variable that takes on the values 0 and 1 only; used in one form of *multiple correspondence analysis* to code multivariate categorical data.

- *eigenvalue* — a quantity inherent in a square matrix, part of a decomposition of the matrix into the product of simpler matrices; in general, a square matrix has as many eigenvalues and associated eigenvectors as its rank; in the context of CA, eigenvalue is a synonym for *principal inertia*.

- *Euclidean distance* — distance measure between vectors where squared differences between corresponding elements are summed, followed by taking the square root of this sum.

- *identification condition* — a condition which needs to be imposed on an optimization problem in order to obtain a unique solution.

- *independence model* — (also called the "homogeneity hypothesis") a model for the counts in a *contingency table*, which assumes that the rows (or columns) are sampled randomly from the same population; i.e., the expected relative frequencies (proportions) in each row, or in each column, are the same.

- *indicator matrix* — the coding of a multivariate categorical data set in the form of *dummy variables*.

- *inertia* — weighted sum of squared distances of a set of points to their *centroid*; in CA the points are *profiles*, weights are the *masses* of the profiles and the distances are *chi-square distances*.

- *interactive coding* — the formation of a single categorical variable from all the category combinations of two categorical variables.

- *joint correspondence analysis (JCA)* — an adaptation of *multiple correspondence analysis* to analyse all unique two-way cross-tabulations of a set of Q categorical variables while ignoring the cross-tabulations of each variable with itself.

- *map* — a spatial representation of points (row and column profiles in CA) with a distance or scalar product (*biplot*) interpretation.

- *mass* — the marginal total of a row or a column of a table, divided by the grand total of the table; used as weights in CA.

- *multiple correspondence analysis (MCA)* — for more than two categorical variables, the CA of the *indicator matrix* or *Burt matrix* formed from the variables.

- *optimal scale* — a set of scale values assigned to the categories of several categorical variables, which optimizes some criterion such as maximum correlation (with another variable) or maximum discrimination (between a set of groups).

- *outlier* — a point on the periphery of a display that is well separated from the general scatter of points.

- *partial bootstrap* — in CA, the display of many replicate samples, obtained by *bootstrapping*, as supplementary points in the map of the original table.

- *permutation test* — generation of data permutations, either all possible ones or a large random sample, assuming a null hypothesis, in order to obtain the null distribution of a test statistic and thus estimate the P-value associated with the observed value of the statistic.

- *principal axis* — a direction of spread of points in multidimensional space that optimizes the *inertia* displayed; can be thought of equivalently as an axis which best fits the points in a weighted least-squares sense.

- *principal coordinates* — coordinates of a set of points projected onto a *principal axis*, such that their weighted sum of squares along an axis equals the *principal inertia* on that axis.

- *principal inertia* — the *inertia* displayed along a *principal axis*; also referred to as an *eigenvalue*.

- *profile* — a row or a column of a table divided by its total; the profiles are the points visualized in CA.

- *scalar product* — for two point vectors, the product of their lengths multiplied by the cosine of the angle between them; directly proportional to the projection of one point on the vector defined by the other.

- *simplex* — a triangle in two dimensions, a tetrahedron in three dimensions, and generalizations of these geometric figures in higher dimensions; in CA

J-dimensional *profiles* lie inside a simplex defined by J *vertices* in $(J-1)$-dimensional space.

- *singular value decomposition (SVD)* — a matrix decomposition similar to that of *eigenvalues* and eigenvectors, but applicable to rectangular matrices; the squares of the singular values are eigenvalues of particular square matrices, and the left and right singular vectors are also eigenvectors.

- *skew-symmetric matrix* — a square matrix with zeros on the diagonal and the property that the elements above the diagonal have the same absolute value as those opposite them below the diagonal, but with opposite sign.

- *stacked tables* — a number of contingency tables, usually based on cross-tabulating the same individuals, concatenated row-wise or column-wise or both.

- *standard coordinates* — coordinates of a set of points such that their weighted sum of squares along an axis equals 1.

- *subset correspondence analysis* — a variant of CA which allows subsets of rows and/or columns to be analysed, while maintaining the same geometry of the full table.

- *supplementary point* — a point which has a position (a *profile* in CA) with mass set equal to zero; in other words a supplementary point is displayed on the map but has not been used in the construction of the map.

- *transition relationship* — the relationship between the row and column co-ordinates in a map.

- *vertex* — a unit profile, i.e. a profile with all elements zero except one with value 1.

- *Ward clustering* — a specific hierarchical clustering algorithm which minimizes the within-cluster inertia at each clustering step, equivalent to maximizing the between-cluster inertia.

- *weighted Euclidean distance* — similar to *Euclidean distance*, but with a positive weighting factor for each squared difference term.

Epilogue

Correspondence analysis (CA) has been presented in this book as a versatile method of data visualization, applicable in a wide variety of situations. This epilogue serves to elaborate further on certain aspects of the method that arise frequently in discussions, and to add some personal thoughts.

The interpretation of the symmetric map remains one of the method's most controversial aspects, even though it is the option of choice for CA maps. This map displays both rows and columns in principal coordinates — that is, the projections of the row profiles and the projections of the column profiles are shown in a joint map even though they occupy different spaces. We have seen (see, for example, Chapters 9 and 10) that the difference between the symmetric map and the asymmetric map (where the points do lie in the same space) is the rescaling along principal axes by the square root of the respective principal inertias. Thus the directions indicated by the points in principal coordinates and by their counterparts in standard coordinates are almost the same when the square roots of the principal inertias are not too different — this can be seen clearly in Exhibit 13.4 where the biplot axes, which pass through the vertex points, almost coincide with the corresponding profile points. In such a case, the biplot style of interpreting the display is valid whether the display is symmetric or asymmetric. If the square roots of the principal inertias are very different, however, there can be problems with the biplot style interpretation of the symmetric map — see, for example, the differences in the directions defined by the smoking categories in Exhibits 9.2 and 9.5. Even so, the distortion induced by using the symmetric map as if it were a true biplot is not so great, as discussed in the following paper by Gabriel:

The symmetric map

—Gabriel, R. (2002). Goodness of fit of biplots and correspondence analysis. *Biometrika*, **89**, 423–436.

This means that the scaling debate is really an academic issue and, as far as the practice of CA is concerned, hardly worth all the discussion that it has generated. In my opinion, the symmetric map is still the best default map to use, and is the default option in our **ca** package for R. If the data matrix is to be interpreted asymmetrically, with the rows (say) representing "observational units" (e.g., demographic groups in sociology such as marital status and educational levels, or sampling locations in ecology or archeology, or texts in linguistics, etc.) and the columns representing "variables" (e.g., response categories in sociology, species in ecology, artefacts in archeology, or stylistic indicators in linguistics, etc.), then the standard CA biplot is a good alternative, since it displays optimally the distances between the units

267

and gives a valid biplot interpretation of the units projected onto the variable directions, as well as giving a meaningful length to the variable vectors.

"You can't have your cake and eat it, too!" This English saying is unfortunately true in this case, as well as the similar expression *"You can't get everything in life!"* It would be wonderful if we could represent optimally in a single map all three things we would like to interpret:

i. the distances between the row profiles,

ii. the distances between the column profiles, and

iii. the scalar products between row and column points, which reconstruct the original data (i.e., the biplot).

But the reality is that we can see at most only two of these three represented optimally at the same time. In the symmetric map we see optimal representations of the chi-square distances for the row profiles and for the column profiles; hence row-to-row distances and column-to-column distances can be interpreted (i.e., (i) and (ii)). The row–column relationship is not optimally represented, but can still be interpreted with reasonable assurance, taking into account the remarks on the previous page. In the asymmetric map we see the optimal representation of one set of profiles, say the row profiles, while the column vertices give the extreme profiles as reference points and also lie on the biplot axes for interpreting the optimal row–column relationship (i.e., (i) and (iii)). The standard CA biplot is a variation of the asymmetric map which also shows one set of profiles, say the row profiles, but pulls in the column vertices by the square roots of their masses to improve the joint representation (i.e., (i) and (iii)). In this biplot the lengths of the column vectors on the biplot axes can be related to their contributions on the principal axes (see Chapter 13).

"Symmetrical normalization" in SPSS Apart from the free R and commercial XLSTAT options described in the Computational Appendix, we have not discussed other software packages which include CA, among which are Minitab, Stata, Statistica, SPAD, SAS and SPSS. Because SPSS is widely used, a comment about its options is necessary here. In SPSS's CA program in the *Categories* module, an alternative biplot is given that has not been illustrated in this book, called the "symmetrical normalization", which may be confused with the symmetric map described in this book. It is not exactly the same thing, however, since it uses standard coordinates scaled by the square roots of the singular values (i.e., fourth roots of the principal inertias) instead of the singular values themselves: in other words (referring to (A.8) and (A.9) on page 202), $\mathbf{\Phi D}_\alpha^{\frac{1}{2}}$ and $\mathbf{\Gamma D}_\alpha^{\frac{1}{2}}$ instead of the symmetric map's $\mathbf{\Phi D}_\alpha$ and $\mathbf{\Gamma D}_\alpha$. SPSS's "symmetrical normalization" gives optimal representation of scalar products but non-optimal representations of distances since neither rows nor columns are represented in principal coordinates. Hence this display gives only one of the three things we want

(i.e., (iii) but not (i) and (ii)). Even though the difference between this display and the symmetric map is also a matter of scale factors along the two axes, and in most cases is hardly distinguishable to an untrained eye, we would not recommend this map in practice since it represents no benefit (in fact, a loss) over existing options. If the principal inertias on the two axes are fairly close, then, as before, the relative positions of points in the "symmetrical normalization" is practically identical to those in the symmetric map, but the symmetric map is definitely preferable since it shows the chi-square distances to their true scale. For purposes of comparison, this option is provided in our R package **ca**, where it is called the "symmetric biplot", specified as follows in the plotting: `map="symbiplot"` (see page 234). Curiously, the symmetric map, one of the most popular display options of French researchers, has not been available in previous versions of SPSS, and it is still not possible in the latest versions to obtain a joint map of the rows and columns in principal coordinates. The best one can do is to select the "principal" normalization, which gives the row and column principal coordinates in numerical form, but the program stubbornly refuses to make a joint map of them, preferring separate maps. Unless the user's raw respondent-level data are in SPSS format, the CA program in SPSS is not recommended. However, the other optimal scaling programs in *Categories*, for multiple correspondence analysis (called by its synonym, homogeneity analysis, in previous versions) and nonlinear principal component analysis (*CatPCA*) are very useful for social science applications.

The issue of rare categories and their effect on the χ^2-distance and the CA solution is also one that has generated much discussion, especially in ecological circles, almost entirely without justification. For example, C. R. Rao has stated that "since the chi-square distance uses the marginal proportions in the denominator, undue emphasis is given to the categories with low frequencies in measuring affinities between profiles" — see page 42 of the following article:

Rare (low-frequency) categories

— Rao. C.R. (1995). A review of canonical coordinates and an alternative to correspondence analysis using Hellinger distance. *Qüestiió*, **19**, 23–63. Download:

www.idescat.net/sort/questiio/questiiopdf/19.1,2,3.1.radhakrishna.pdf

But the fact is that in CA each category is weighted in the analysis proportional to its mass, which reduces the role played by low-frequency categories in the analysis. It can be shown very simply by looking at the numerical contributions of each category to the principal axes that rare categories generally have low influence in the solution — i.e., the solution would be almost the same if they were removed from the analysis entirely.

As an illustration, for the species abundance data set of Chapter 10 (see page 77), we calculated the relative abundance for the 10 most abundant and 10 least abundant species and compared this to their respective percentage contributions to the first two axes of the CA map in Exhibit 10.5. The results are as follows:

Species	Relative abundance	Contributions to axes Axis 1	Axis 2
10 most abundant	74.6%	77.3%	89.3%
10 least abundant	0.4%	0.8%	0.5%

This illustrates that the rare species do not make an excessive contribution to the two-dimensional solution — the contributions are very much in line with the abundances in each subset of species. Only a few times, in our experience, do low-frequency categories make an excessively large contribution to the major principal axes, in which case they should be removed or combined with another category. A case in point is in sociological applications, when low frequency categories such as missing values coincide in the same subgroup of respondents. These categories can dominate an MCA solution, often defining the first principal axis as in Exhibits 18.2 and 18.5. This situation can be rectified by using a subset analysis, or by combining rare response categories with others in a sensible way. The analogous situation in ecology would be when several rare species co-occur in the same samples, but this is not a common situation — usually rare species occur randomly in different samples.

Low-frequency categories are often outliers

Having said that the rows or columns with low frequencies generally have low influence on the solution, because of low mass, it is true that these points are often outliers in the CA map, owing to their strange profiles. Outliers draw attention to themselves and it is probably for this reason that the impression is given that they may be affecting the analysis strongly. As shown in Chapter 13 and mentioned above, the standard CA biplot would solve this problem by "pulling in" these points by the square root of their masses, which effectively eliminates the low frequency outliers since these end up close to the origin. This also demonstrates graphically that their influence on each principal axis is quite low.

χ^2-distance is a Mahalanobis distance

This section is a bit technical but will demonstrate to the statistically minded reader that the chi-square distance, apart from being the key to all the properties of CA, can also be defended on theoretical grounds as an appropriate statistical distance measure. A bit of matrix notation is needed for the weighted Euclidean distance function in (5.1), which can be written as:

$$\text{weighted Euclidean distance} = \sqrt{(\mathbf{x} - \mathbf{y})^{\mathsf{T}} \mathbf{D}_w (\mathbf{x} - \mathbf{y})} \qquad (E.1)$$

where $\mathbf{x}$ and $\mathbf{y}$ are vectors with elements x_j and y_j, $j=1,\ldots,J$, ^{T} indicates transposition of a vector or matrix, and $\mathbf{D}_w$ is the diagonal matrix of the dimension weighting factors w_j's. The rows, say, of a contingency table can be assumed to be realizations of a *multinomial* random variable. The multinomial distribution is a generalization of the binomial distribution, and is a model to describe the behaviour of data sampled from a population where there are probabilities p_j, $j = 1, \ldots, J$ of observing a sampling unit in one of J groups, for example our three readership groups in Chapter 3 (see Exhibit 3.1 on page 18). Under the null hypothesis that the data are sampled from the same population, the five education groups in this data set would be multinomial samples from a population with probabilities p_1, p_2, p_3, where the estimates of p_j for the three groups are the elements of the average profile $\hat{p}_1 = c_1 = 0.183$, $\hat{p}_2 = c_2 = 0.413$ and $\hat{p}_3 = c_3 = 0.404$ (see last row of Exhibit 3.1). The classic distance function for grouped multivariate data is the so-called *Mahalanobis distance*, based on the inverse of the covariance matrix of the variables:

$$\text{Mahalanobis distance} = \sqrt{(\mathbf{x} - \mathbf{y})^{\mathsf{T}}\mathbf{\Sigma}^{-1}(\mathbf{x} - \mathbf{y})} \tag{E.2}$$

which looks like the weighted Euclidean distance (E.1), except that it involves a full square matrix of weights $\mathbf{\Sigma}^{-1}$, not a diagonal matrix. The covariance matrix $\mathbf{\Sigma}$ for the multinomial distribution has a simple form, for example for our trinomial case $J = 3$ (the results are similar for any number of groups):

$$\mathbf{\Sigma} = \begin{bmatrix} p_1(1-p_1) & -p_1p_2 & -p_1p_3 \\ -p_2p_1 & p_2(1-p_2) & -p_2p_3 \\ -p_3p_1 & -p_3p_2 & p_3(1-p_3) \end{bmatrix} = \mathbf{D}_p - \mathbf{p}\mathbf{p}^{\mathsf{T}} \tag{E.3}$$

where $\mathbf{p}$ is the vector of the p_j's and $\mathbf{D}_p$ is the corresponding diagonal matrix. (E.3) is estimated by substituting the probabilities p_j by their estimates c_j. To invert the covariance matrix $\mathbf{\Sigma}$ in the usual way is not possible since it is a singular matrix, so we cannot find a matrix $\mathbf{\Sigma}^{-1}$ such that $\mathbf{\Sigma}\mathbf{\Sigma}^{-1} = \mathbf{I}$. One way to get around this is to drop one of the categories and use just $J-1$ categories throughout. Whichever category is omitted, the Mahalanobis distance will be the same. An alternative more elegant approach, which is entirely equivalent but uses all J categories, is to use a so-called *generalized inverse*, denoted by $\mathbf{\Sigma}^-$, which has the property that $\mathbf{\Sigma}\mathbf{\Sigma}^-\mathbf{\Sigma}=\mathbf{\Sigma}$ (this is also known as the *Moore-Penrose inverse*). It turns out that the Moore-Penrose generalized inverse of (E.3) is equal to:

$$\mathbf{\Sigma}^- = \begin{bmatrix} 1/p_1 & 0 & 0 \\ 0 & 1/p_2 & 0 \\ 0 & 0 & 1/p_3 \end{bmatrix} = \mathbf{D}_p^{-1} \tag{E.4}$$

which means that the Mahalanobis distance in (E.2) is estimated exactly by the χ^2-distance. The situation here is similar to that in linear discriminant analysis: to discriminate maximally between groups, the groups are assumed to have equal covariance matrices, which in the multinomial case translates to our assuming the independence model, and then the data vectors are embedded in Mahalanobis space, which translates to chi-square space.

The issue of rotations has not been treated in this book because seldom are rotations justified or needed in CA. On the one hand, the profile space is not unbounded real vector space but a space delimited by the unit points, or vertices, defining a simplex in multidimensional space. The idea of lining up the category points along specific axes at right angles does not have the same meaning as in factor analysis where right-angledness really means zero correlation between variables (remember that one category point in CA is always determined by all the others, because the elements of a profile add up to 1). Rotations can be appropriate in some contexts in MCA and nonlinear PCA (not treated in this book) where several variables are analyzed simultaneously. For example, it frequently occurs that all non-response points in MCA lie together in a bunch owing to high association in the data set but not coinciding with a principal axis, in which case it would be good to be able to rotate the solution as a way to "partial out" the non-response points. But this problem can be better solved by doing a subset analysis (Chapter 21) which completely ignores the non-response points and focuses totally on the substantive responses. If rotation of a solution is required, then the masses of the category points should be taken into account in the rotation. For example, a weighted version of the usual varimax rotation in factor analysis would be to maximize the criterion (assuming rotation of the column points is required):

$$\sum_j \sum_k c_j^2 (\tilde{y}_{jk}^2 - \frac{1}{J} \sum_{j'} \tilde{y}_{j'k}^2)^2 \tag{E.5}$$

where $\tilde{y}_{jk}$ is the rotated standard coordinate, that is the (j, k)-th element of $\hat{\mathbf{Y}} = \mathbf{YQ}$, for $\mathbf{Q}$ an orthogonal rotation matrix. Notice that the mass c_j is squared because the objective function involves the fourth powers of the coordinates. Since $c_j \tilde{y}_{jk}^2 = (c_j^{\frac{1}{2}} \tilde{y}_{jk})^2$, an almost identical alternative is suggested, which is a small modification of the usual varimax criterion: perform a rotation (unweighted) on the rescaled standard coordinates $c_j^{\frac{1}{2}} y_{jk}$, which are exactly those used in the standard CA biplot. In other words, rotate the solution to concentrate (or *reify* in factor analysis terminology) the contributions of the categories on the rotated axes.

CA and modelling

In Chapter 13 CA in K^* dimensions was shown to be a decomposition which can be written as follows (see (13.4), also (A.14) in the Theoretical Appendix):

$$p_{ij} = r_i c_j + r_i c_j \left(\sum_{k=1}^{K^*} \sqrt{\lambda_k} \phi_{ik} \gamma_{jk} \right) + e_{ij} \quad i = 1, \ldots, I; \ j = 1, \ldots, J \tag{E.6}$$

The CA solution is obtained by minimizing the weighted sum of squares of the residuals e_{ij}. The first part of the decomposition, $r_i c_j$, is the expected value under the model of independence, so that the second part is explaining the deviations from the independence model as the sum of K^* bilinear terms (this bilinear part has a geometric interpretation in K^* dimensions which is the subject of most of this book). Any other model of the user's choice can be substituted for the independence model. For example, in the following article, the authors consider log-linear models for a contingency table, and then use CA as a way of exploring the structure, if any, in the deviations from the log-linear model:

— van der Heijden, P.G.M., de Falguerolles, A., and de Leeuw, J. (1989). A combined approach to contingency table analysis and log-linear analysis (with discussion). *Applied Statistics*, **38**, 249–292.

This strategy can be used for multiway tables as well, using a contingency table modelling approach to account for main effects and chosen interactions in a first step, then calculating the residuals from the model and analyzing these by CA. But note that this is not a straightforward application of CA, since the data have already been centred with respect to the model. The centring step in CA must not be performed and the original margins of the table must be used in the weighted least-squares fitting.

CA has a close affinity to *spectral mapping*, a method developed originally by Paul Lewi in the 1970s and used extensively in the analysis of biological activity spectra in the development of new drugs. A more recent reference is:

—Lewi, P.J. (1998) Analysis of contingency tables. In *Handbook of Chemo-metrics and Qualimetrics: Part B* (eds. B.G.M. Vandeginste, D.L. Massart, L.M.C. Buydens, S. de Jong, P.J. Lewi, J. Smeyers-Verbeke), Chapter 32, pp. 161–206. Amsterdam: Elsevier.

CA and spectral mapping

Spectral mapping operates on the logarithms of the table, but incorporates the same weighting of rows and columns as in CA, i.e., by the row and column masses computed on the original table. The log-transformed table is double-centred with respect to the weighted row and column averages before applying the SVD as in CA. If the inertia in the data is low, then spectral mapping and CA are almost identical. The difference in the two methods is greater when the inertia are higher. Spectral mapping involves mapping the logarithms of ratios of the data, and has very interesting model-diagnostic properties. It also obeys the principle of distributional equivalence (see pages 37–38) and, in addition, has the property of *subcompositional coherence*, which is the property which underpins the analysis of compositional data: since the ratios between two data values remains the same whether or not rows or columns are excluded from the table, subsets of rows or columns can be analyzed with impunity. In CA on the other hand, profiles and distances are affected when analyzing subsets; that is, CA is not subcompositionally coherent, hence the special adaptation called subset CA described in Chapter 21. For more details and further references, consult the following unpublished working paper:

—Greenacre, M.J. and Lewi, P.J. (2005). Distributional equivalence and sub-compositional coherence in the analysis of contingency tables, ratio-scale measurements and compositional data. Working Paper no. 908, Department of Economics and Business, Universitat Pompeu Fabra, Barcelona. Download: www.econ.upf.edu/en/research/onepaper.php?id=908

To conclude this epilogue, here is an unsolved problem. We know that in simple CA the dimensionality of an $I \times J$ table is $\min(I-1, J-1)$. For a $J \times J$ Burt matrix based on Q categorical variables, the dimensionality is $J - Q$, but we know that $J - Q$ dimensions are much more than we need to reproduce the off-diagonal tables exactly. We propose that the dimensionality of a Q-variable data set be defined as the number of dimensions required to reproduce

The dimensionality of a multivariate categorical data set

the $\frac{1}{2}Q(Q-1)$ cross-tabulations exactly. In other words, the dimensionality is the number of dimensions required in a joint CA to explain 100% of the inertia. The question is: can this dimensionality be determined beforehand or does it need to be discovered empirically? Using the rule in adjusted MCA of considering only the K^* dimensions for which $\sqrt{\lambda_k} > 1/Q$ (see page 149, for example the adjusted inertia in (19.7)), it would be convenient if this provided the clue to the dimensionality. In empirical studies the inertia explained using this number (K^*) of dimensions is usually very close to 100%, but this is no proof, of course, that the dimensionality is K^*. Perhaps, by the time a third edition of this book is published, this problem will have been solved!

Index